BOTANICAL RESEARCH AND PRACTICES

THE REPRODUCTION STRATEGY OF PAPHIOPEDILUM ARMENIACUM (ORCHIDACEAE)

BOTANICAL RESEARCH AND PRACTICES

Additional books in this series can be found on Nova's website at:

https://www.novapublishers.com/catalog/index.php?cPath=23_29&seriesp=Botanical+Research+and+Practices

Additional e-books in this series can be found on Nova's website at:

https://www.novapublishers.com/catalog/index.php?cPath=23_29&seriespe=Botanical+Research+and+Practices

BOTANICAL RESEARCH AND PRACTICES

THE REPRODUCTION STRATEGY OF PAPHIOPEDILUM ARMENIACUM (ORCHIDACEAE)

ZHONGJIAN LIU
LIJUN CHEN
AND
QING ZHOU

Novinka
Nova Science Publishers, Inc.
New York

For permission to use material from this book please contact us:
Telephone 631-231-7269; Fax 631-231-8175
Web Site: http://www.novapublishers.com

LIBRARY OF CONGRESS CATALOGING-IN-PUBLICATION DATA

Available Upon Request

ISBN: 978-1-61668-205-7

Published by Nova Science Publishers, Inc. † New York

CONTENTS

PREFACE

This book studies and discusses the relations among the three fundamental life activities: growth, maintenance and reproduction in *Paphiopedilum armeniacum* (Orchidaceae), with an emphasis on plant propagation. The biological and ecological characteristics of *Paphiopedilum armeniacum* both *in situ* and *ex situ*, the population performance in different vegetation patterns, the asexual reproductive mechanism and the occurrence and maintenance of sexual reproductive mechanism including selfing, crossing and mix-breeding in flowers of population with deceptive pollination have been thoroughly studied in this book.

Chapter 1 – Breeding is the core of the evolutionary process of plants, of which the basic life procedures include growing, surviving and breeding. The selection on reproductive evolution in plants is powered by their adaptation to the habitat. The choices of habitat in plants have developed into various behaviours of clonal reproduction and sexual reproduction, but it is the sexual reproduction and its crossing advantages and inbreeding depression that usually attract attention, while the fact that the plant serves its existence and multiplication better via the reproductive assurance of clonal reproduction and selfing is neglected. *Paphiopedilum armeniacum* grows in limestone area and has both asexual and sexual reproduction. It is an ideal plant for study on plant survival and propagation.

Chapter 2 – *Paphiopedilum armeniacum* is an endangered orchid species, endemic to China. During the period of April 2000 to October 2007, tests of the biological and ecological characteristics of *P. armeniacum* and its *ex situ* reproductive modes had been carried out in Nushan Mountain, Yunnan Province, to study the relationships between the climate, vegetation and other environmental factors in original habitat and the growing, surviving and breeding of *P. armeniacum*. The orchid grows and survives well in limestone areas. It has very strong abilities of asexual and sexual reproduction, and it is endowed with the specialties of highly effectively using its environment and tolerating bad conditions, and with strategies of extending plant lifespan and escaping death, as well as the ability to control the ageing process. *Paphiopedilum armeniacum* is able to invade a forest environment with moderate demolishment and those in the early period of restoration but cannot live in large dense forests.

Chapter 3 – The population performances of *P. armeniacum* under different floristic composition patterns were compared to determine how abundance of this species correlates with the floristic composition pattern. Based on the floristic composition, four end groups of plots within this vegetation at level 3 of the TWINSPAN program can be classified. The DCA ordination of plots data showed that the variation in the floristic composition of four groups appeared to be related to the disturbance degree, the succession stage and the ground covering. There were probably two kinds of succession series of population performances: one was at the site where it was covered with shallow soil and rocky chipping and the performance of population showed a trend from colonization stage to expansion stage; the other was at the site where it was covered with deeper soil and the population development exhibited a slight downtrend. The difference of population performance based on different succession series further indicated that the *P. armeniacum* was a semi-shaded species with low competitive abilities in well-developed soil, and the reproduction of this species was restricted by the growing condition that was considered to be of optimization.

Chapter 4 – *Paphiopedilum armeniacum* has two modes of asexual reproduction—by tillering and by producing rhizomes. Tillering of asexual reproduction is of strong vitality and is endowed with an emergency mechanism to cope with adversity by producing numerous rhizomes. The purpose of clonal reproduction is to extend the lifespan of genets, to implement the sexual reproduction and to increase the proportion of sexual mating. The rhizome reproduction in *P. armeniacum* is also an adaptation to the condition of defoliation or to the condition covered with soil in order to escape from death. *Paphiopedilum armeniacum* still keeps the original reproductive pattern in *ex situ* habitat with an obviously different condition from the original one. Clonal genets also keep the patterns of growing, surviving and breeding, which are adaptive to the uncertain conditions after returning to the original habitat. All further indicate that its reproductive mode has evolved from ecological adaptation to genetic adaptation.

Chapter 5 – *Paphiopedilum armeniacum* is able to produce many seeds through sexual reproduction and the seeds can grow into new genets effectively. It deceives various pollinators among populations to achieve selfing, crossing, or even mixed mating in flowers by mimicking flowers of associated plants. In conditions of unsaturated pollinators or external pollen, self-pollination completed by pollinators could increase seeds production. The high inbreeding depression in the self-pollinating estimate of this study showed that, as the pollinating environment turned into an essential factor of reproductive assurance for wild species, the increase of the selfing rate could increase the effect of reproductive assurance. We demonstrated how the visiting behavior of pollinating insects in a changeable pollinating environment influenced the reproductive assurance. This demonstration indicates that flowers with a strict cross-pollinating structure also have a selfing mechanism and a changeable pollinating environment is propitious to the selfing and the mixed mating, which associates the reproductive assurance with the selfing rate of flowers of non-autonomous self-pollination via the mechanism of mixed mating to display the diversity of mixed mating evolution. The mix-breeding mode of *P. armeniacum* undounbtedly reflects the success in evolution of flowering plants in terms of the diversity of breeding strategies, and it also gives an insight into why the vast majority of plants have self-compatibility, or most plants employ reproductive strategies of selfing or mixed mating, although crossing is of superiority.

Chapter 6 – This chapter summarizes the studies of the above five chapters. It is believed that *Paphiopedilum armeniacum* makes the best of environment to serve propagation so that it gains relative competitive advantage, which represents that the change of adaptability related to the life history evolution may be the essential condition of its being an invasive species. It indicates that plant not only has the ability to develop crossing but also has the ability to promote selfing and employ selfing, and the sexual change may be the product from the tradeoff in life history of new habitat invasion and habitat fluctuation. The revelation gained from the reproductive strategy of *P. armeniacum* is that, there is no plant can use any environment, and few environments are not used by plant, and then gives insight into the protection of this species.

Chapter 1

THE PROPAGATION OF PLANT

ABSTRACT

Breeding is the core of the evolutionary process of plants, of which the basic life procedures include growing, surviving and breeding. The selection on reproductive evolution in plants is powered by their adaptation to the habitat. The choices of habitat in plants have developed into various behaviours of clonal reproduction and sexual reproduction, but it is the sexual reproduction and its crossing advantages and inbreeding depression that usually attract attention, while the fact that the plant serves its existence and multiplication better via the reproductive assurance of clonal reproduction and selfing is neglected. *Paphiopedilum armeniacum* grows in limestone area and has both asexual and sexual reproduction. It is an ideal plant for study on plant survival and propagation.

A Darwinian demon, that is that every plant would hold a limitlessly long lifespan, start its propagation directly after being born and produce numerous seeds every time, is an organism that does not exist in the real world (Partridge and Harvey, 1988), and there is no probability that each seed can grow into a new individual. That's because the amount of resources for any plant at any time is always limited. Every living creature could only find a best plan of utilization for resources and energy amongst any possible combinations with growth, survival and reproduction, in order to achieve its reproductive success to the utmost extent. Therefore, to study the breeding of plants, it is compulsory to study how plants try to find a best plan for reproduction to carry out effective allocation of resources in those conflicted demands, the most important one of which is to be reflected in the breeding system of plants. The reproductive ways of flowering plants and their mating countermeasures are extremely complicated, and the breeding system usually represents all the characteristics that directly influence the genetic components of descendants (Wyatt, 1983). Such characteristics include the integrated characteristics of the flower, the lifespan of every floral reproductive organ, the flowering patterns, the degree of self-compatibility and the mating system, at the same time they also include the behaviour of pollinators (Holsinger, 2000) and the clonal reproduction of non-sexual expression during the reproductive process (Liu et al., 2006, 2007). So, to understand the highly complicated diversities of mating and reproduction of plants, it is necessary to study the basic life activities of plants in growth, survival and reproduction, in order to determine the selective mechanisms of evolution and preservation of breeding characteristics and the consequence of inbreeding fitness (Barrett, 1998; Liu et al., 2006).

Regarding the evolution of the mating system, when the selection is able to effectively identify and preserve the homozygous genotype with higher fitness, selfing can immediately increase the fitness of the plant's population and enhance the adaptability of the population to its local environment (Stebbins, 1957; Grant, 1958; Baker, 1959). Such increase of fitness is considered to be short-term and it is at the cost of decreased adaptability to future changes of the environment (Darlington, 1939; Mather, 1943; Wang and Dang, 1999). However, the trade-off between short-term fitness and long-term flexibility of evolution is a characteristic at the population level. Whether selfing can evolve or not compulsorily depends on whether it is good for individuals, rather than depending on the advantage or disadvantage that are judged at the population level (Holsinger, 2000). Despite the theory being widely accepted that selfing is considered as harmful to populations owing to the general existence of the inbreeding depression, in fact, the individual of selfing is usually selected because it has a higher fitness among the crossing populations (Lloyd, 1979; Liu et al., 2008).

But why can selfing bring widespread attention? It is because firstly, the relative frequency of selfing has a great influence on the fitness of plant individuals via the effects of inbreeding depression and the advantage of hybridization; secondly, the relative frequency of selfing and crossing is the most important decisive factor in the genetic structure of plant populations, which can influence the genetic diversity within the population as well as the genetic specialization among populations; finally, selfing has greatly changed the morphological features of flowers and the countermeasure of resource allocation (Barrett et al., 1996). So studies of the mating system in plants are mainly focused on the evolution of the selfing rate. Darwin (1859) had already noticed that selfing could result in the falloff in fertility and in the vitality of seeds, and he also put forward that the reason why selfing could evolve was because it could assure the reproduction of plants in conditions that the external pollen is insufficient. After that many studies have given additional evidences to the advantages of selfing (Wang and Dang, 1999; Liu et al., 2007, 2008). Therefore, as for studies on the mating system, close attention should not only be paid to how floral characteristics influence the behaviour of pollinators, the transferring efficiency of pollen and the mating partners (Harder and Barrett, 1996), but also paid to the male fitness, that is the gene amount passed down by pollen, in what way from the plant; as well as to how selfing affects the survival of plants and the development of plant communities. Only if the genes that can facilitate the sefling of the plant have been encountered with strong depression from negative selection, will the plant be able to invade into the population soon; it is because those individuals from selfing could pass down more genes to the next generation than those from crossing (Fisher, 1941). If the internal relation between selfing and crossing is neglected, and the mode and efficiency of pollen transfer, the influence of the floral mechanism on the mating system and the ecological process that includes pollination have not been given enough attention, the study on breeding is not comprehensive. It is because the mating opportunity is depended on the transfer of pollen and the mating pattern is not completely in agreement with the pollination pattern. Therefore, the understanding of the transfer patterns of pollen is important to reveal the evolutionary trend of mating system in flowering plants, which demands the comprehension of floral pattern, growth characteristics and pollinators behaviour that can influence self-pollination. So, it's important to emphasize the significance of inbreeding depression, as well as the function of pollination details before fertilization. The selfing rate of an individual plant not only depends on the amount of pollen from self-pollination, but also on how much external pollen it receives from another individuals and the occasion when it receives, as well as the relative abilities of competion between selfing pollen and crossing pollen to catch the opportunity of fertilization (Lloyd, 1979, 1992; Holsinger, 1991, 1996). The choice of pollinating strategy of a plant is dependent on its habitat, and a special geographic environment can bring out special species and its breeding strategy, especially the strategy of self-pollination (Liu et al., 2008). And it needs an integrated knowledge of pollination biology and mating system to overall and correctly understand the ecology and evolution of plant breeding. Inbreeding depression caused by self-pollination is considered as one of the selective pressures on the evolution of the breeding system in plants

(Darwin, 1876; Charlesworth and Charlesworth, 1987, 1990; Lande and Schemske, 1985; Holsinger, 1988, 1991; Li et al., 2002). And out-crossing is considered to be the main motive force of pollination diversity in the breeding system. However, it is a paradox that selfing is the reproductive strategy that a majority of angiosperms will choose (Barrett, 2002; Lloyd and Schoen, 1992; Liu et al., 2007, 2008).

What on earth has driven them to make such a choice? However, most researches on the pattern of mating evolution are mainly focused on the function of inbreeding depression or on the structure of communities that decide the balanced level of selfing (Kalisz et al., 1999).

In fact selfing has many advantages, such as, invading a new habitat (Baker, 1955; Stebbins, 1957), avoiding unreliable pollinators (Darwin, 1876; Lloyd, 1979; Piper et al., 1984), maintaining the adaptability of the population to its local habitat (Antonovics, 1968) and automatically passing genes down to the next generation (Jain, 1976).

All of those advantages impel flowering plants to bring about many selfing mechanisms, for instance, self-pollination achieved by anther self-transfer (Liu et al., 2006) and geitonogamy (Liu et al., 2008), which has always been neglected in the research of breeding systems (de Jong et al., 1999; de Jong, 2000).

Moreover, the knowledge of its ecology is extremely deficient (Liu et al., 2006; Snow et al., 1996). If a plant population relies only on animal agents to spread pollen among individuals, there will be a selective predominance on individuals that can self-pollinate but have not crossed previously under the condition that there are no pollinators or inadequate pollinators (Lloyd, 1979).

But why plants still choose self-pollination when there are animal agents, and even the selective predominance for self-pollination only takes place in the presence of animal agents? Selfing in higher plants is attributed to ecological factors, the foremost of which is the variation of pollinators during seasons and years.

This probably neglects the potential capability of automatically meeting an emergency of plant itself to adapt to new habitats and the capability of actively making use of insects' behaviour to better serve the reproductive strategy in plants (Liu et al., 2008).

Therefore, it requires us to investigate the evolutionary significance of selfing mechanism from the aspects of ecology and phylogeny of plants. At least we should not ignore the positive effect of selfing on the evolution of plants in terms of studies of breeding systems, nor isolate the influences on sexual reproduction from clonal reproduction of non-sexual expression, and from growth and survival during the breeding process.

This new species, *Paphiopedilum armeniacum*, published by Chen Singchi and Liu Fanghuan (1982), arouses great concerns in horticulture and around the world.

Paphiopedilum armeniacum is a terrestrial or lithophytic plant, often with several to many creeping stolons over 10 cm long and 2–3 mm thick. Leaves 4–7, oblong or narrow-oblong, thick-coriaceous, 5–18 cm long, 1.5–3.5 cm wide, acute or also tridenticulate at apex, adaxially tessellated with dark and light green, abaxially densely purple-spotted and carinate, serrulate-margined. Scape erect, 10–28 cm long, pale yellow-green densely spotted with purple, pubescent; bract ovate-lanceolate or ovate, 1.4–1.8 cm long, pale yellow-green spotted with purple, puberulous; pedicel and ovary 2–4 cm long, white-pubescent; flower single, 5–9 (–11) cm across, yellow, pale yellow-green or golden yellow; sepals and petals sometimes slightly flushed with pale green toward their abaxial apexes; staminode with pale maroon veins; dorsal sepal ovate, 2.1–4.8 cm long, 1.4–2.5 cm wide, subacute at apex, pubescent on both surfaces; synsepal similar to dorsal sepal, slightly smaller, obtusely bicarinate abaxially; petals broadly ovate-elliptic to suborbicular, 2.8–5.3 cm long, 2.5–4.8 cm wide, subacute or rounded at apex, slightly undulate-margined, ciliate, white-villose toward adaxial base; lip deep pouched, subglobose, 4–6 cm long, 2.5–4 cm wide, with involute margins apically; staminode broadly ovate or ovate-orbicular, 1.2–2.2 cm long and wide, acute at apex, margins recurved. $2n=26$. (Figure 1.1).

When *P. armeniacum* had its first exhibition in America in 1983, all people were deeply impressed by its magnificent and stunning flower and the judges gave it the highest record-breaking score (92) in all prizes of American Orchid Society.

Thereafter, *P. armeniacum* and its variations and hybrids totally won 71 awards from American Orchid Society by 1992 and the highest score was up to 97. It's the precedent in the history of orchid (Liu et al., 2009).

This study chooses *Paphiopedilum armeniacum*, a perennial herbaceous orchid with great ornamental value, as the research material to study its reproductive strategy and its characteristics: distributing in limestone area, opening a single flower on each genet, special floral structure, spatial isolation between pistils and stamens, possessing both asexual and sexual reproduction, and producing a great amount of seeds.

Figure 1.1. The morphology of *Paphiopedilum armeniacum.*

References

Antonovics, J. (1968). Evolution in closely adjacent plant populations. V. Evolution of self-fertility. *Heredity*, *23*, 219–238.

Baker, H. G. (1955). Self-compatibility and establishment after "long distance" dispersal. *Evolution*, *9*, 347–349.

Baker, H. G. (1959). Reproductive methods as factors in speciation in flowering plants. *Cold Spring Harbor Symposium on Quantitative Biology*, *24*, 177–191.

Barrett, S. C. H. (1998). The evolution of mating strategies in flowering plants. *Trends in Plant Science*, *3*, 335–341.

Barrett, S. C. H. (2002). The evolution of plant sexual diversity. *Nature Reviews Genetics*, *3*, 274–284.

Barrett, S. C. H., Harder, L. D., Worley, A. C. (1996). The comparative biology of pollination and mating in flowering plants. *Philosophical Transactions of the Royal Society of London Series B-Biological Sciences, 351*, 1271–1280.

Charlesworth, D. and Charlesworth, B. (1987). Inbreeding depression and its evolutionary consequences. *Annual Review of Ecology and Systematics*, *18*, 237–268.

Charlesworth, D. and Charlesworth, B. (1990). Inbreeding depression with heterozygote advantage and its effect on selection for modifiers changing the outcrossing rate. *Evolution*, *44*, 870–888.

Chen, S. C. and Liu, F. Y. (1982). Notes on some species of *Paphiopedilum* from Yunnan. *Acta Botanica Yunnan*, *4* (2), 163–167.

Darlington, C. D. (1939). *Evolution of genetic systems*. Cambridge: Cambridge University Press.

Darwin, C. (1859). *On the Origin of Species by Means of Natural Selection or the Preservation of Favored Races in the Struggle for Life*. London: John Murray.

Darwin, C. (1876). *The Effects of Cross and Self Fertilization in the Vegetable Kingdom*. London: John Murray.

de Jong, T. J. (2000). From pollen dynamics to adaptive dynamics. *Plant Species Biology*, *15*, 31–41.

de Jong, T. J., Klinkhamer, P. G. L., Rademaker, M. C. J. (1999). How geitonogamous selfing affects sex allocation in hermaphrodite plants. *Journal of Evolutionary Biology*, *12*, 166–176.

Fisher, R. A. (1941). Average excess and average effect of a gene substitution. *Annals of Eugenics*, II: 53–63.

Grant, V. (1958). The regulation of recombination in plants. *Cold Spring Harbor Symposium on Quantitative Biology*, *231*, 337–363.

Harder, L. D. and Barrett, S. C. H. (1996). Pollen dispersal and mating patterns in animal-pollinated plants. In: Lloyd D. G. and Barrett S. C. H., ed. *Floral biology: studies on floral evolution in animal-pollinated plants*. New York: Chapman and Hall, 140 –190.

Holsinger, K. E. (1988). Inbreeding depression doesn't matter: the genetic basis of mating system evolution. *Evolution*, *42*, 1235–1244.

Holsinger, K. E. (1991). Inbreeding depression and the evolution of plant mating systems. *Trends in Ecology and Evolution*, *6*, 307–308.

Holsinger, K. E. (1996). Pollination biology and the evolution of mating systems in flowering plants. *Evolutionary Biology*, *29*,107–149.

Holsinger, K. E. (2000). Reproductive systems and evolution in vascular plants. *Proceedings of the National Academy of Sciences, USA*, *97*, 7037–7042.

Jain, S. K. (1976). The evolution of inbreeding in plants. *Annual Review of Ecology and Systematics*, *7*, 469–495.

Kalisz, S., Vogler, D., Fails, B., et al. (1999). The mechanism of delayed selfing in *Collinsia verna* (Scrophulariaceae). *American Journal of Botany*, *86*, 1239–1247.

Lande, R. and Schemske, D. W. (1985). The evolution of self-fertilization and inbreeding depression in plants. I. Genetic models. *Evolution*, *39*, 24–40.

Li, Q. J., Kress, W. J., Xu, Z. F., et al. (2002). Mating system and stigmatic behavior during flowering of *Alpinia kwangsiensis* (Zingiberaceae). *Plant Systematics and Evolution*, *232*, 123–132.

Liu, Z .J., Chen, L. J., Lei, S. P., et al. (2007). The reproduction strategy of *Trias verrucosa* from China. *Acta Ecologica Sinica*, *27* (11), 4460–4468.

Liu, Z. J., Chen, L. J., Liu, K. W., et al. (2008). *Chenorchis*: a new orchid genus and its eco-strategy of ant pollination. *Acta Ecologica Sinica*, *26* (6), 2433–2444.

Liu, Z. J., Chen, S. C., Chen, L. J., Lei, S. P. (2009). The Genus *Paphiopedilum* in China. Beijing, Science Press.

Liu, Z. J., Liu, K. W., Chen, L. J., et al. (2006). Conservation ecology of endangered species *Paphiopedilum armeniacum* (Orchidaceae). *Acta Ecologica Sinica*, *9*, 2791–2800.

Lloyd, D. G. and Schoen, D. J. (1992). Self- and cross-fertilization in plants. I. Functional dimensions. *International Journal of Plant Sciences*, *153*, 358–369.

Lloyd, D. G. (1979). Some reproductive factors affecting the selection of self-fertilization in plants. *American Naturalist*, *113*, 67–79.

Lloyd, D. G. (1992). Self- and cross-fertilization in plants. II. The selection of self-fertilization. *International Journal of Plant Science*, *153*, 370–380.

Mather, K. (1943). Polygenic inheritance and natural selection. *Biological Reviews*, *18*, 32–64.

Partridge, L. and Harvey, P. H. (1988). The ecological context of life-history evolution. *Science*, *241*, 1449–1455.

Piper, J. G., Charlesworth, B., Charlesworth, D. (1984). A high rate of self-fertilization and increased seed fertility of homostyle primroses. *Nature*, *310*, 50–51.

Snow, A. A., Spira, T. P., Simpson, R., et al. (1996). The ecology of geitonogamous pollination. In: Lloyd D. G, Barrett S. C. H., ed. *Floral Biology: Studies on Floral Evolution in Animal-pollinated Plants*. New York: Chapman and Hall.

Stebbins, G. L. (1957). Self-fertilization and population variability in the higher plants. *American Naturalist*, *91*, 337–354.

Wang, C. Y. and Dang, C. L. (1999). Plant mating system and its evolutionary mechanism in relation to population adaptation. *Journal of Wuhan Botanical Research*, *17*, 163–172.

Wyatt, R. (1983). Pollinator-plant interactions and the evolution of breeding systems. In: Real, L., ed. *Pollination biology*. Orlando: Academy Press, 51–95.

Chapter 2

THE GROWTH AND SURVIVAL OF PAPHIOPEDILUM ARMENIACUM

ABSTRACT

Paphiopedilum armeniacum is an endangered orchid species, endemic to China. During the period of April 2000 to October 2007, tests of the biological and ecological characteristics of *P. armeniacum* and its *ex situ* reproductive modes had been carried out in Nushan Mountain, Yunnan Province, to study the relationships between the climate, vegetation and other environmental factors in original habitat and the growing, surviving and breeding of *P. armeniacum*. The orchid grows and survives well in limestone areas. It has very strong abilities of asexual and sexual reproduction, and it is endowed with the specialties of highly effectively using its environment and tolerating bad conditions, and with strategies of extending plant lifespan and escaping death, as well as the ability to control the ageing process. *Paphiopedilum armeniacum* is able to invade a forest environment with moderate demolishment and those in the early period of restoration but cannot live in large dense forests.

Most growing plants have to complete their pollination, depending on an 'intermediary agent' to achieve the mating of both gender gemets. Such a kind of dependance results in many pollination mechanisms via intermediary agents to spread pollen, such as secreting nectar to attract pollinators, but there is also the breeding mechanism of asexual reproduction with no need for the mating of gender gemets or that of autonomous self-pollination not depending on any agents (Liu et al., 2006). Breeding is the core of the evolutionary process, as well as the key in any study of biological evolution. And the description on any breeding system of plants at least includes the following three aspects: 1) whether the population is of asexual reproduction or sexual reproduction; 2) if it is of sexual reproduction, what the floral gender characteristics (gender system) are/is; and 3) what are the floral characters that influence selfing and crossing (Silvertown and Charlesworth, 2001).

The amount of genes passed down to the descendants from each plant of sexual reproduction either via seeds or via pollen must be the same on the level of population. But there may be differences among different individuals within a population: some individuals pass more genes to the next generation via seeds than via pollen; and others may transfer genes mainly via pollen and they do not fruit or seldom fruit. Therefore, the production of seed cannot represent all the effects of breeding of each individual. To increase the female fitness of a plant by avoiding selfing is considered to be the main or even the unique selective power of floral characters. And the breeding history of flowering plants, to a greater extent, is the record of how plants avoid selfing in different ways (Darlington, 1963). In recent years, people have gradually realized that the knowledge of breeding biology of plants gained from the aspect of the production of seeds is incomplete, and they have started to do more and more work trying to re-define the adaptive meaning of floral characters in the light of avoiding the mutual interference of male and female functions and increasing the male fitness as a result (Lloyd and Yates, 1982; Lloyd and Webb, 1986, Bertin, 1993; Bertin and Newman, 1993; Barrett, 1998, 2002a, b, 2003). That whether plants facilitate selfing in any different ways and why they facilitate selfing are questions worthy of further investigation, which are exactly relative with the growth and survival of plants. In order to study the specific process and details of a plant's breeding and to examine the procedure of sexual breeding, it is necessary to take the influence of its biology into consideration, and integrate the procedure of sexual breeding that authentically unite the two main subbranches of the evolutionary ecology: the evolution of life history and the breeding ecology, rather than only study the breeding system of plants that mainly investigates the genetic matters of mating patterns concerned with the inbreeding depression. First of all, we should pay attention to the allocation of resources among the three activities of basic life: growing, surviving and breeding.

Paphiopedilum armeniacum Chen et Liu is a *Paphiopedilum* species that has been found recently (Chen and Liu, 1982; Chen, 1996, 1999; Liu et al., 2009), and it is an attractive and popular plant and has already won the highest Golden Awards and First Class Certificate (FCC) of many international flower exhibitions and orchid exhibitions (Chen and Luo, 2003). *Paphiopedilum armeniacum* is a species endemic to China which attracts international attention, however, owing to the excessive collection, rampant smuggling for export, and the destruction of habitat, the quantity of this beautiful orchid has decreased rapidly in the recent decades, and it has even been on the verge of extinction (Liu et al., 2004, 2009). Therefore, it is necessary and highly significant to study its distribution and the structure of its community in order to find out what type of ecological environment is needed for this species to maintain its existence and the best habitat it needs in different stages during its life history, so as to provide evidences for further studies on its breeding ecology.

1. Geographic Location and Climate of Sites for Observation and Experiment

The site for study on breeding ecology of *P. armeniacum* is located in Luoshapo Mountain, Laowo Village of Lushui County, Yunnan Province (99°03′E, 25°54′N), China, where the type of climate is a subtropical upland monsoon. The site for *ex situ* conservation is at the National Orchid Conservation Center in Wutongshan Mountain in Shenzhen, Guangdong, China (114°10′E, 22°35′N), with the subtropical ocean climate where the perennially dominant wind comes from the southeast, and this location is often affected by typhoons in summer. The climatic parameters of observed site (original habitat) and *ex situ* site are summarized in Table 2.1 (Zhao, 1999; Chen, 2003).

Table 2.1. Climate parameters of original and *ex situ* sites

Site	Alt. (m)	Average tem. (℃)			Extreme tem. (℃)		Annual Frost-free (d)	Annual rain (mm)	Less rain season	rainy season	Annual sunshine (h)	High tem. season	Low tem. season
		Annual	Coldest month	Hottest month	Max.	Min							
In situ	1700–2150	15.1	9.1	19.6	31.8	0.6	282	1213	Nov. – May	Jun. – Oct.	2045	Jul. – Aug.	Dec. – Feb.
Ex situ	30–50	22.4	14.1	28.2	38.7	0.2	355	1883	Oct. – Mar.	Apr. – Sep.	2120	Jun. – Sep.	Dec. – Feb.

During the observation in Luoshapo Mountain, the record of daytime temperature was from 10℃ (at 6:00) to 46℃ (at 16:00, the temperature of earth's surface under sunlight), and the humidity was from 12% (daytime) to 100% (at night). It snows in winter from time to time. The pedogenic parental rock in this site is mainly composed of weathered limestone with limestone gravel in 1 – 2cm diameters everywhere except for a little earth on the shallow surface layer, and the soil is barren. The gradient is 60 –70° (90°). The air around the bottom of the mountain is heated by the sunshine and then it rises, forming strong hill wind.

2. METHODS

2.1. Geographic Distribution

From October 1996 to September 2004, based on documents reviewing and locals inquiring, the investigative team composed of technicians and local people visited Gaoligongshan Mountain, Nushan Mountain and Yunling of Hengduanshan Mountain in Western Yunnan, China, where were located with the known and possible sites of *P. armeniacum*, so that we can draw the outline of distribution range of this species after investigation.

2.2. Biological Characteristics

2.2.1. Observation of Habitat And Growing Condition

From April to June 2000, 40 sampled sites (1 m × 1 m each) were randomly selected for observation of habitat and growing condition in the growing area of *P. armeniacum* in Luoshapo Mountain, to carry out the following experiments: count the genets (ramets) of each site; observe the habitat (including the canopy density of community, the accosiated plants, and the micro landform, etc.); record the growing condition of each gemet (soil condition), and the growing situation (with or without environmental stress), and the extension of roots (depth, direction and width). Specimens of plants in each site were collected for identification and analyses.

2.2.2. Observation on Flowering Condition

From May to June 2001, 8 sampled sites (2 m × 2 m each), located from low altitude to high altitude (ranging from 1700 m to 2150 m) with a separation of 50 m at altitude were selected and observed for 3 times: the first time was in the flowering period; the second after the florescence; and the third during the fruit-ripening period. And the following informations were recorded each time: the clonal form and size of each genet, and the flowering status and the quantity of fruits that could successfully produce seeds after natural fertilization of each genet. From May to September 2004, the marked flowering ramets were randomly selected for observation to verify the above results.

2.2.3. Detection on Pollen Histochemistry

Matured pollen grains of 10 freshly openning flowers with dehisced anthers or with mucus were collected at random and smeared on the slide for observation. The pollen samples were immersed in IKI solution or Sudan IV solution, and examined under a microscope. A blue or black colour (stained with IKI) indicates the presence of starch, and a red colour (stained with Sudan IV) shows the presence of lipids (Dafni, 1992). Experiments of each flower were repeated twice and the corresponding colour was recorded.

2.2.4. Detection on Effects of Both Stigma and Stamens Upon the Flowering Duration

In order to examine effects of both stigma and stamens upon the flowering duration, the following treatments were set up: a) bagged before flowering and removed stigma after flowering, then bagged again; b) bagged before flowering and removed stamens after flowering, then bagged again; c) bagged before flowering and removed both the stigma and stamens after flowering, then bagged again; d) bagged before flowering without any treatment after flowering (natural control).

2.3. Observation on Life Cycle

In May 2001, 10 fruited genets were randomly selected from Luoshapo Mountain and 10 sampled plots (50 cm × 50 cm each) were set with each genet in the center ($n = 10$).

The soil was dug loosely to check carefully and confirm that there was no protocorm, and then refilled with humous. After fruits had split, the quantity of germinating seeds, and the processes of growth and development, blossom, fruiting, tillering and withering of *P. armeniacum* were observed and recorded to ascertain its life cycle.

3. Results and Analyses

3.1. Geographic Distribution

After investigation, the geographic distribution of *P. armeniacum* has its principal area in the mid section of Nushan Mountain Range, extending to the east coast of Lancangjiang River and partial area of the west coast of Nujiang River (Figure 2.1).

3.2. Biological Characteristics

3.2.1. Vegetation and the Structure of Community

The vegetation of study site consists of secondary broadleaf shrubberies and tussocks, and the vegetation is seperated by limestones forming a fasciculate distribution. The shrubberies are 1–1.5 m tall with a canopy density of 70% – 90%.

As being divided by rocks and slope, the surface among forests can be shined by the setting sun. Plants in the study site include shrubbery: *Campylotropis capillipes, Castanopsis delavayi, Deutzia monbeigii, Excoecaria acerifolia, Hypericum beanii, Inula cappa, Itea yunnanensis, Leptodermis forrestii, Osyris wightiana, Photinia prionophylla, Pistacia weinmannifolia, Quercus spinosa, Rhododendron rubiginosum, Viburnum mullaha, Vitis henryana* etc; herbs: *Carex lanucisquamata, Carex pergracilis, Ficus tikoua, Gerbera delavayi, Leptoboea multiflora, Leucas martinicensis, Shuteria involucrata* var. *glabrata, Reineckea carnea* etc; Orchidaceae: *Cymbidium faberi* Rolfe, *C. tortisepalum* Fukuyama, *C. serratum* Schltr., *C. cyperifolium* var. *szechuanicum* S. C. Chen et Z. J. Liu, *Cypripedium plectrochilum* Franch, etc.

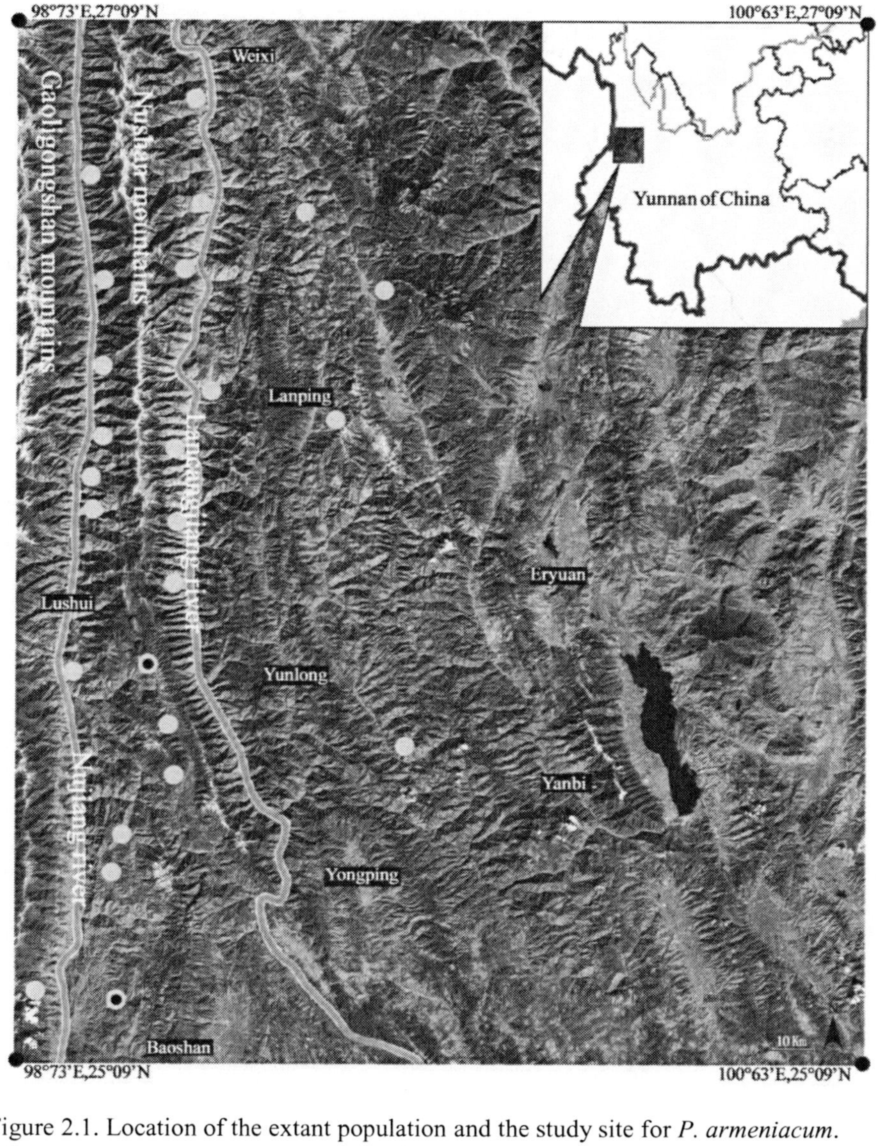

Figure 2.1. Location of the extant population and the study site for *P. armeniacum*. ●: locations of the extant population; ●: location of the study site.

3.2.2. Growing Condition

Paphiopedilum armeniacum mostly grows on the northeast of mountain, entirely terrestrial and has its blocky distributions in bosk, tussock, bare slope, and rock gap, and it is a kind of gregarious plants. *Paphiopedilum armeniacum* has more plants in the community with herbaceous plants such as *Quercus*, *Eremopogon delavayi* (Hack.) and *H. beanni* growing nearby or more on the surface of rocks which is near these plants. This orchid can grow in forest with a canopy density of 0 – 90% and it can endure strongly direct sunlight and grow in forest with steep slope of high canopy density on which the setting sunlight can reach. However, it has never been found in dense forest of large arbor trees or place of moist and flat lands with deep soils and seeper or mountains with an altitude higher than 2250 m (Figure 2.2).

In Gaoligongshan Mountain closed to the west coast of Nujiang River, *P. armeniacum* has only been found growing in the side of large rocks and place among rock cluster where no large tree grows.

The shrubbery in this area is deciduous alpine plant, xerophil, and with thick cutin on leaves, which will turn into humus and whose hard-to-decay veins and spines can keep the humus loose in a long term. The herbaceous plants can fix the humus with their roots and hold the seeds flowing along with water, so that the seeds can germinate in the rich humus and grow into seedlings.

Figure 2.2. Habitat of *P. armeniacum* in Nushan Mountain. Here, low hill areas are exploited into farmland and *P. armeniacum* loses its growing space; and the relatively stable place in high hill areas is affected by overgrazing; while there are many plants on surface of rocks in steep slope, of which posterity may rapidly invade artificially sparse forest and artificial blanks.

Figure 2.3. *Paphiopedilum armeniacum* is able to grow in inclement environment. This picture shows that a plant grows in the aperture of limestone with its scape outstretching to flower, waiting for the visit of insects.

Therefore, *P. armeniacum* mostly grows on rock surfaces with more humus in bosks or tussocks and place of steep slope where a rivulet is blocked. Some *P. armeniacum* can grow on and along the nearly upright crags or in shallow cave of limestone and some grow on bare limestones with little humus in the lacunose surface (Figure 2.3).

Paphiopedilum armeniacum ramets growing in the same microhabitat have no obvious difference in growing condition and size, but there are some differences in the amount of individuals in different habitats. The leaves of genets growing on the surface of bare rocks or on crags will become yellowish and wizened for being short of water in dry season, and even in severe conditions some leaves even dry out or they will be burned on the leaf apex. However, no plants died from drought or were found damaged because of snow or frost.

The roots of *P. armeniacum* spread shallowly, and mainly grow in humus on crushed rocks, about 2–3 cm deep and no more than 5 cm. The extension of roots in transverse direction is wider but the extending degree is non-uniform. The roots develope better and the extension is more widely on the side with richer humus, about 2–3 times wider than the coverage area of plant. Even so,

P. armeniacum can be easily pulled out because of its weak adherence to the ground.

Paphiopedilum armeniacum has capabilities of self-reintroduction into the nature and self-rejuvenation. For instance, there are mass *P. armeniacum* plants growing on the surface of an artificial slope and trench (at an altitude of 1900m), which was dug in 1984 for preparing to build a hydraulic power station, and many well-developed small *P. armeniacum* communities on the remnant walls of an abandoned temple at the altitude of 2130 m. And there are not any other plants growing in such communities. In addition, there is also a great deal of *P. armeniacum* genets in sparsely artificial fir forest and artificial tussocks, and these communities have already completed alternation of several generations.

3.2.3. Flowering Status

In the 8 selected sites along the gradient of altitude, the flowering status of totally 1302 ramets of 443 genets of *P. armeniacum* were observed. The ramets in the sampled sites started their flowering season in late April, and the full florescence was in early May and the flowering season finished in early June. Fruits matured from middle September to early October. The duration of flowering without insect pollination was (31.2 ± 4.7) days (n = 30). The flowering status of each site of the 3 observations on 10 May, 10 June and 8 October 2001 is shown in Table 2.2.

86 sampled flowers were randomly selected and marked on 1 May 2004, 32 flowers of which were fertilized and fructified till 16 June, taking up 37.21% of the marked flowers, and the other 54 flowers all dried out and fell off; by 20 September there were 28 matured fruits, equivalent to 32.56% of the marked flowers. These results were similar to those from the 8 sampled sites selected along the gradient of altitude.

The flowering season of *P. armeniacum* will change markedly, which depends largely on the arrival time of rainy season: *P. armeniacum* germinates and forms buds in autumn and during dry winter the buds will remain in dormancy until the coming of rainy season.

The rain will break the dormancy of buds and brings up the flowering season. If there is a drought in spring, the flowering season will be delayed. Genets of *P. armeniacum* can flower even though producing only one ramet. Each *P. armeniacum* ramet can only flower once during its life cycle and does not germinate any more after its first germination, regardless of whether the first germination leads to successful flowering or not.

Table 2.2. Blooming state of *P. armeniacum* at eight sites

Site no.	Alt. (m)	Quantity		Quantity of blooming or fruiting ramet		Quantity of failed blooming ramet			Σ	Rate of blooming (%)	Fruit quantity	Fruit set rate (%)
01	1700	52	146	2	3	3	1	1	10	6.85	5	50.00
02	1750	55	163	2	2	4	2	2	12	7.36	6	50.00
03	1800	56	171	3	4	4	1	2	14	8.19	4	28.57
04	1850	59	178	4	3	4	2	1	14	7.87	5	35.71
05	1900	51	149	3	2	2	1	0	8	5.37	2	25.00
06	1950	53	145	2	3	3	1	1	10	6.90	2	20.00
07	2000	62	186	4	2	5	2	2	15	8.06	3	20.00
08	2150	55	164	6	4	4	0	0	14	8.54	4	28.57
Σ		443	1302	26	23	29	10	9	97	59.14	31	257.85
$\bar{x}$		55.38	162.75	3.25	2.88	3.63	1.25	1.13	12.13	7.39	3.88	32.23
SD		3.66	15.25	1.39	0.83	0.92	0.71	0.83	2.53	1.02	1.46	12.08

3.2.4. Pollen Histochemistry

After detection, no colour change occurred in the pollen samples after IKI treatment, that is, it remained the original colour and didn't turn into blue or black, but a red colour merged after Sudan Ⅲ treatment, showing the existence of lipids in pollen.

3.2.5. The Influence of Both Stigma and Stamens on Flowering Duration

The result of experiments is presented in Table 2.3. In the treatments of only removing the stigma and both removing the stigma and stamens, the flowering duration (till the corolla falls off from the ovary) was both 7±1.3d; the flowering duration of removing stigma was significantly different from the flowering duration of only removing stamens (35.45±7.4d) (t=9.87624, P<0.01, n=60); the flowering duration of removing stamens was obviously different from that of bagged without manipulation (27.7±3d) (t=2.58561, P<0.01, n=60); the flowering duration of removing stigma had no distinguished difference with those of self-pollination (7±1.9d) and cross-pollination (7±0.7d), (t = 0.6038, P >0.05, n = 60; t = 0.5843, P > 0.05, n = 60).

Table 2.3. Comparison of flowering time at different experimental pollen and stigma treatments (mean±s.d.)

Treatments	Quantity of flowers	Flowering time (d)		
		Mean	SD	CV
Stigma removed in bud (bagged)	60	7	1.3	0.1856
Anthers and stigma removed in bud (bagged)	60	7	1.3	0.1856
Anthers removed in bud (bagged)	60	35.45	7.4	0.3273
No treatment in bud (bagged)	60	27.7	3	0.1215
Selfed (bagged)	60	7	1.9	0.2714
Outcrossed (bagged)	60	7	0.7	0.1000

The result of experiments indicated that the existence of stigma and the accomplishment of its function were highly associated with the flowering duration, and the flowering duration had been extended obviously if the stamens were removed.

3.3. Life Cycle

During the late September 2001, all fruits of the 10 marked plants had split and scattered seeds. When checked in June 2002, the seeds had grown into protocorms of 1.5 cm in diameter, and the protocorms produced underground stems; in July 2002, buds on the apex of underground stems started to grow leaves and they had 2–3 leaves till October 2002. On an average, there were 2.3 ± 1.3 seedlings (n = 10) in each sampled plots. When checked in May 2003, each seedling had set up 4–6 leaves and 8 plants of them had been flowering from May to June 2004, 3 of which had fruited. From May to June of 2005, 15 plants had been flowering and 6 of them had fruited. All ramets started to germinate tillering buds after flowering season and fruited period. Under the artificial culture conditions, some strong genets could also grow 1–2 tillering bud/buds from the base of ramet after germinating buds.

It will take about 1–4 years for *P. armeniacum* genet to complete its life cycle from buds germination to all leaves aging, leaving the remnant stem; but there will be great differences among different individuals. When the tillering bud emerges from the mother ramet in spring, it can grow 4–5 leaves in the same year, but some new tillering buds cannot flower normally when there is damage on the apex, such as, insect bite, but they will grow new tillering buds in the following autumn. Under general circumstance, *P. armeniacum* can germinate tillering buds only once a year, and when it has 4–5 leaves or more, it will stop its vegetative growth and shift to the differentiation of alabastrum. Ramets will flower after growing 1–2 years, and give off new tillering buds and then they will gradually senesce and wither away, but the underground stems can still survive for years which could germinate tillering buds once more and grow into new genets after artificial division. The quantity of tillering buds and the senescence of ramets after sexual reproduction have relation to the microenvironment: under favorable conditions, ramets can grow up to 4 generations, whereas under unfavorable conditions, they will wither after only one generation. A genet generally grows 2 or 3 clonal ramets.

4. Discussion

Paphiopedilum armeniacum is endowed with biocological substructure of enduring extreme environment. It can grow in environments of great range of variation between high temperature (46℃) and low temperature (0.2–0.6℃), raininess and drought, high altitude and low altitude, reflecting that this species has a great adaptability to climate and habitat and it can act as an important role in the ecological evolution of *in situ* habitat. *Paphiopedilum armeniacum* is a kind of vanguard plants for vegetation recovery; it can resuscitate synchronously with the vegetation recovery when the destruction in habitat is stopped, and it even can rehabilitate its growth faster than other plants. *Paphiopedilum armeniacum* cannot grow well in good woody environments but it grows largely in secondary forests, man-made forests, made-made slash, meadow, and even on the surface of bare rocks, which indicates that *P. armeniacum* has some weed-like characteristics and its growing environment needs to be disturbed somewhat, which is a kind of distinctiveness in the biology of this species.

Based on the life cycle of *P. armeniacum*, it does not require a long time for the new individuals of *P. armeniacum* to grow from seeds in original habitat to complete the subrogation of generations, and it doesn't require any special skills to invade new habitat. Thus, it is believed that *P. armeniacum* is a kind of dissimilative species during its process of evolution and it adapts well to worsened environment and even depends on such environment. Therefore, moderate deterioration or destruction in environment is even good for its multiplication, but the continued degeneration of environment that exceeds its endurance will inevitably result in its extinction. On the other hand, if plant communities in the distribution range have evolved into that as communities in dense forest of large trees in Gaoligong Mountains on the western Nujiang River, it would become extinct because of its terrestrial short plants which could not twine around the trees like *P. villosum* (Cribb, 1998; Liu et al.,2009). Of course, it can hardly evolve into large forest in mountains of bare limestones and cliffy slope, which may be the reason why *P. armeniacum* can survive and has its narrow distribution in this region.

Another survival reaction of *P. armeniacum* is the self-controlled ability of aging process after the fructification of flowering ramets. In favorable environment, ramets will dry rot after growing 4 years, whereas in unfavorable environment, the mother genet will begin to age when germinating new tillering buds after fructification, obviously in order to support the growth of new plants with its limited resource, which could avoid the conflict between parental generations and filial generations (Hardling and Nilsson, 2001). This self-controlled ability is a trade-off in the life history of *P. armeniacum* (Obeso, 2002).

From the view of flowering status, *P. armeniacum* is endowed with highly effective asexual reproduction owing to the unsaturation of sexual reproduction. Its flower doesn't have any structural defects in sexual reproduction and this species cannot produce asexual seeds. It is the habitat that brings up the breeding strategy of asexual reproduction for the purpose of its survival and propagation, which is an adaptation of this species to different habitat by accumulating resource to produce perennial plants, in order to ensure its occupation in the habitat. It may be an evolutional strategy of this species in environments of drought, great difference in temperature, strong wind, strong sunlight and lack of pollinating insects (Liu et al., 2007). In order to adapt to the rhythmly change of climate, *P. armeniacum* develops a relevant growing rhythm, which is an adaptation to the resource utilization of surrounding environment in terms of physiology and configuration that is reflected by its phenology. As a perennial plant, the beginning time of its growth and procreation and their duration all have relation to the local climatical condition. The characteristic that it can survive for years after flowering and its thicker leather leaves both acclimatize itself to the local ecological conditions, which are in favor of increasing its fitness in habitat of drought, cold, high temperature and leanness. On the other hand, it is also reflected that this species has the ability of producing mass clonal ramets to invade new habitat or avoid unfavorable habitat (Liu et al., 2007). *Paphiopedilum armeniacum* itself has both asexual reproduction and sexual reproduction. Asexual reproduction can grow many tillering buds and possesses the emergent mechanism of producing mass rhizomes to cope with harmfulness; effective pollinating insects in the habitat can make some flowering plants fruited, and there are abundant seeds in a fruit and quite a few seeds can effectively germinate into new genets. At the same time, *P. armeniacum* is endowed with characteristics of highly utilizing environment and tolerating harsh environment, and strategies of extending genet lifespan and escaping from death and the controlling ability of aging process. This is one reason why *P. armeniacum* can grow well and adapt well to the local climatic environment of limestone area in Nushan Mountain, Yunnan, China.

Many researches have indicated that pollen grains in all angiosperm contain lipid, but starch dosen't always exist and the existence of starch has relation with the ecology and the classification, as well as the phylogeny (Baker and Baker, 1979). It is generally believed that possessing amylaceous pollen is the characteristic of anemophilous plants, and as for anthophilous plants, there is more or less lipid in pollen. Pollen grains of some species were partially or completely served as rewards for visitors, such as bees and flies, and the natural selection usually is in favor of pollen grains with lipid; in contrast, the natural selection would favor amylaceous pollen in case the pollen grains do not serve as nutriment for insects. In contrast, as for pollen grains of some species that do not serve as rewards for visitors, the natural selection is in favor of pollen grains with starch. The result of observation indicated that the pollination of *P. armeniacum* is entomophilous pollination and it is completed by bees and hoverfly; even though its pollen grains contain lipid as those of associated plants, they do not serve as reward substance for bees or hoverfly. This may be an evolutionary characteristic of plants with function of deceiving pollinators.

The flowering of plant is for the breeding of offsprings, and the plant would end its flowering as soon as its purpose is achieved or unable to achieve, which is an evolutionary adaptation. The result of test on the effect of both stigma and stamen on flowering showed that the existence of stigma and its function are the key to maintain the opening of flowers; once the function has been finished or the stigma is disappeared, it is meaningless for the flower to continue its opening, so it has to end its opening in order to save resource. The stamens of cross-pollination plants are prepared for other flowers, so the removal of stamens is similar to pollen grains taken away by pollinators, which dose not have any influence on the opening of flowers. This is also a successful strategy of pollination during evolution in flowers of *P. armeniacum*. The pollen grains are able to keep their viability for a long time after departing from the flower, which is beneficial for the long-distance pollination of pollinating insects and suffice the communication of genes among populations; meanwhile, the stigma is able to receive pollen grains from different flowers during its receptive period, which could bring up diversity of its progeny. However, removing the stamen could obviously extend the flowering duration, it is possible because the removal of stamens reduces the resource expense of flowers, or the stamens could produce some substance that lengthens the opening of flower, however, this phenomenon still needs further study.

REFERENCES

Baker, H. G. and Baker, I. (1979). Starch in angiosperm pollen grains and its evolutionary significance. *American Journal of Botany*, *66*, 591–600.

Barrett, S. C. H. (1998). The evolution of mating strategies in flowering plants. *Trends in Plant Science*, *3*, 335–341.

Barrett, S. C. H. (2002a). The evolution of plant sexual diversity. *Nature Reviews Genetics*, *3*, 274–284.

Barrett, S. C. H. (2002b). Sexual interference of the floral kind. *Heredity*, *88*, 154–159.

Barrett, S. C. H. (2003). Mating strategies in flowering plants: the outcrossing-selfing paradigm and beyond. *Philosophical Transactions of the Royal Society of London Series B-Biological Sciences*, *358*, 991–1004.

Bertin, R. (1993). Incidence of monoecy and dichogamy in relation to self-fertilization in angiosperms. *American Journal of Botany, 80*, 557–560.

Bertin, R. I. and Newman, C. M. (1993). Dichogamy in Angiosperms. *The Botanical Review*, *59*,112–152.

Chen, L. E. ed. (2003). *Plants in Wutong Mountain*. Beijing: China Forestry Publishing House, 1.

Chen, S. C. and Liu, F. Y. (1982). Notes on some species of *Paphiopedilum* from Yunnan. *Acta Botanica Yunnan*, *4* (2), 163–167.

Chen, S. C. and Luo, Y. B. (2003). Advances in some plant groups in China I. a retrospect and prospect of Orchidology in China. *Acta Botanica Sinica*, *45*, 2–20.

Chen, S. C. (1996). Orchids and their conservation in China. *Proceedings 5th Asia Pacific Orchid Conference and Show*. Tokyo: All Japan Orchid Society Publishing, 49–52.

Chen, S. C. (1999). *Paphiopedilum. Flora Reipublicae Popularis Sinicae*. Beijing: Science Press, *17*, 52–72.

Cribb, P. (1998). *The Genus Paphiopedilum*, 2nd edition. Kota Kinabalu: Natural History Publications (Borneo), 254–260.

Dafni, A. (1992). *Pollination ecology, the practical approach*. New York: Oxford University Press.

Darlington, C. D. (1963). *Chromosome Botany and the Origins of Cultivated Plants*. London: Allen and Unwin.

Hardling, R. and Nilsson, P. (2001). A model of triploid endosperm evolution driven by parent-offspring conflict. *Oikos*, *92*, 417–423.

Liu, Z .J., Chen, L. J., Lei, S. P., et al. (2007). The reproduction strategy of *Trias verrucosa* from China. *Acta Ecologica Sinica*, *27* (11), 4460–4468.

Liu, Z. J., Chen, S. C., Chen, L. J., et al. (2009). *The Genus Paphiopedilum in China.* Beijing: Science Press.

Liu, Z. J., Liu, K. W., Chen, L. J., et al. (2006). Conservation ecology of endangered species *Paphiopedilum armeniacum* (Orchidaceae). *Acta Ecologica Sinica*, *9*, 2791–2800.

Liu, Z. J., Zhang, J. Y., Ru, Z. Z., et al. (2004). Conservation biology of *Paphiopedilum purpuratum* (Orchidaceae). *Biodiversity Science*, *12* (5), 509–516.

Lloyd, D. G. and Webb, C. J. (1986). The avoidance of interference between the presentation of pollen and stigmas in angiosperms 1. Dichogamy. *New Zealand Journal of Botany*, *24*, 135–162.

Lloyd, D. G. and Yates, J. M. A. (1982). Intra-sexual selection and the segregation of pollen and stigmas in hermaphrodite plants, Exemplified by *Wahlenbergia Albomarginata* (Campanulaceae). *Evolution*, *36*, 903–913.

Obeso, J. R. (2002). The costs of reproduction in plants. *New Phytologist*, *155*, 321–348.

Silvertown, J. and Charlesworth, D. (2001). *Introduction to plant population biology*, 4th edition. Oxford: Blackwell Science.

Zhao, D. H., ed. (1999). *Map of Yunnan of China*. Beijing: Map press of China, 181–182.

Chapter 3

The Population Performance of Paphiopedilum Armeniacum

Abstract

The population performances of *P. armeniacum* under different floristic composition patterns were compared to determine how abundance of this species correlates with the floristic composition pattern. Based on the floristic composition, four end groups of plots within this vegetation at level 3 of the TWINSPAN program can be classified. The DCA ordination of plots data showed that the variation in the floristic composition of four groups appeared to be related to the disturbance degree, the succession stage and the ground covering. There were probably two kinds of succession series of population performances: one was at the site where it was covered with shallow soil and rocky chipping and the performance of population showed a trend from colonization stage to expansion stage; the other was at the site where it was covered with deeper soil and the population development exhibited a slight downtrend. The difference of population performance based on different succession series further indicated that the *P. armeniacum* was a semi-shaded species with low competitive abilities in well-developed soil, and the reproduction of this species was restricted by the growing condition that was considered to be of optimization.

The relation between density of population and crossing rate is of positive ratio, and that with the selfing rate is of negative ratio. The amount of external pollen received by each plant decreases as the reduction in the amount of potential mates, but the amount of own pollen fallen on the stigma may not be effected by other individuals of the population. In that case, the selfing rate generally decreases as the increase of amount of potential mates (the denseness or size of community) (Farris and Mitton, 1984; Murawski and Hamrick, 1991; Karron et al., 1995; Chen, 2000). Thence, the density of population has influence on the outcome of breeding system in plant and this density is restricted by habitat, in a word, the outcome of breeding is dominated by the population performance.

The response of endangered species to its habitat is an important issue for biologists of conservation (Fahrig, 1997; Webb and Shine, 2000; Ceballos and Ehrlich, 2002; Blomqvist et al., 2003; Watters et al., 2003). The main challenge about this issue is that the relationship between the change in habitat and the decline or increase in communities of different species is unable to determine (Honnay et al., 2002; Ne'eman, 2003; Verheyen and Hermy, 2004; Wotavova et al., 2004; Lin and Xie, 2005). The density of *P. armeniacum* population is highly relative to its habitat and this density is exactly the factor that influences its mating pattern, therefore, to confirm the performance of population in different habitat is the most important thing in studies on breeding of *P. armeniacum*, especially on the relation between the richness of this species and the floristic composition.

1. METHODS

1.1. Observation Site

Based on the investigation of the geographic distribution mentioned above, we selected Laowo village (99°03'E, 25°54'N) of Lushui County, Western Yunnan as our field observation site and the midsection in Nushan Mountain on the south vertical valley of Hengduan Mountains (Figure 2.1; 2.2) as the experiment site to confirm the performance of *P. armeniacum* population.

We visited several mountains (altitude ranges from 1650m to 2350m) near Laowo village where *P. armeniacum* was (or might be) existent. The establishment of observation site was based on the information supplied by local people, for example, Maocao Mountain, Nongjia Mountain, Zhaojia Mountain, Mingchao Mountain, Luoshapo Mountain, ect.. During the investigation, we recorded the presence or absence of *P. armeniacum*, the light condition and the land management of each mountain. We only observed condition of *P. armeniacum* in the northeast Luoshapo Mountain. And then we determined the northeast side of Luoshapo Mountain as the observation site. The terrain of Luoshapo Mountain is rather steep.

1.2. Data Collection

In the northeast side of Luoshapo Mountain, six transverse sections with an area of 2m×50m at the elevation gradient of 50m on the same hillside were selected as observation sites. 10 sampled plots were set on each transverse section at intervals of the same distance. The height of transverse section ranges form 1750 m to 2050 m. The gradient range including the typical concentration of *P. armeniacum* existed. Except the transverse section of 2050 m, 10 sample units were established within each transverse section at an equal distance along transection. Because of the terrain, 9 sample units were established at the transverse section on 2050 m. The area of each sample unit is 4 m^2 (2m×2m). Then each sample unit was subdivided into four equivalent plots (1m×1m). Hence a total of 196 plots were established.

For each unit, we carried out the following assignments: 1) recording the degree of disturbance, types of land coverings (e.g. rock), the percentage of each land coverings, the depth of soil, and the condition of channel and tunnel; 2) estimating the covered percent of all micro-plants in each unit to classify the vegetation; 3) counting the number of *P. armeniacum* individuals and estimating the leaf size of each individual to classify the structure of population. An individual is a visible substantive plant.

1.3. Data Analysis

Vegetation data were analyzed using classification and ordination techniques. TWINSPAN (Two-way Indicator Species Analysis) was used to group the vegetation units according to the pattern of plant composition. To ordinate the vegetation units, we further performed DCA (Detrended Correspondence Analysis) and then qualitatively related the composition pattern to the environmental variables. PC-ORD for Windows, Version 4.20 (McCune and Mefford, 1999) was used for TWINSPAN and DCA using default settings. Based on the result of vegetation analysis, the density, the spatial pattern of distribution, and the size of distribution in the community were estimated in each group. The spatial pattern of distribution was determined via the Variance-to-mean ratio method (Brower et al., 1998). The total size of leaf per individual was used to describe the size of individual.

2. RESULT

2.1. General Characteristics of the Habitat

Among several mountains around Laowo village that we visited, *P. armeniacum* was only found in the northeast side of Luoshapo Mountain. The variation of vegetation is the most visible distinction between Luoshapo Mountain and other mountains. The vegetation of Luoshapo Mountain consists of shrubbery zone occupied by shrubs of 1 – 3 m high or small arbor trees with sparse canopy, whereas the vegetation of other mountains is timberland dominated by trees of 7 – 15 m high with dense canopy. The difference in the degree of disturbance is the main cause of difference in vegetation among these mountains. The observation site in Luoshapo Mountain has experienced excessive deforestation and overgrazing, while other mountains have scarcely been destroyed because they have become parts of the National Ecological Public Benefit Forest Project.

In the northeast side of Luoshapo Mountain, *P. armeniacum* generally rooted in rock rifts with semi-decomposed litter and the basal remnant of stump and few plants were found in sites covered with herbaceous plants where there was deeper soil. And many more plants were found in sites where there was shallow soil or rocky chippings.

2.2. Vegetation Analysis

The condition of the distribution of vegetation is partly summarized in Table 3.1. In a word, there are other 23 species of micro-plants found in the observation site besides *P. armeniacum*, 15 species of which are shrubberies or small arbors and 8 species are herbaceous plant. Based on the floristic composition, four end groups of a unit can be classified at level 3 in TWINSPAN program. The number of units in each group is 32 for group I, 78 for group II, 48 for group III, and 38 for group IV. The species of shrubbery exhibit more variations in the four ending groups, but species of herbaceous plants exhibit fewer variations.

Table 3.1. Species frequency (F)* and dominance (D) in the four end groups of plots based on TWINSPAN**

Species	Group I		Group II		Group III		Group IV	
	F	D	F	D	F	D	F	D
Shrub								
Campylotropis capillipes	III	6.2	V	23.5	IV	5.5	I	0.8
Castanopsis delavayi	-	-	I	0.8	-	-	-	-
Deutzia monbeigii	II	3.6	I	2.6	I	0.9	I	9.7
Excoecaria acerifolia	I	1.2	I	0.1	-	-	-	-
Hypericum beanii	-	-	I	0.2	I	1.1	I	1.6
Inula cappa	I	6.5	I	1.5	I	0.4	-	-
Itea yunnanensis	I	1.0	I	2.4	I	5.2	I	0.6
Leptodermis forrestii	II	3.5	III	5.7	V	26.6	V	46.4
Osyris wightiana	-	-	I	0.4	I	0.1	I	1.6
Photinia prionophylla	-	-	I	1.0	I	1.1	-	-
Pistacia weinmannifolia	-	-	I	0.7	-	-	I	1.7
Quercus spinosa	V	55.2	V	42.9	IV	33.1	IV	25.0
Rhododendron rubiginosum	V	18.5	I	1.5	I	0.9	I	8.4
Viburnum mullaha	II	2.8	II	4.0	I	0.2	I	1.3
Vitis henryana	I	1.7	III	12.9	III	24.8	I	2.9
Herb								
Carex lanucisquamata	V	40.1	IV	35.4	V	33.2	V	36.1
Carex pergracilis	V	55.2	V	63.0	V	63.6	V	58.5
Ficus tikoua	-	-	I	0.6	I	0.6	I	1.7
Gerbera delavayi	I	1.6	I	0.5	I	0.5	I	0.9
Leptoboea multiflora	II	2.9	I	0.9	I	1.1	I	-
Leucas martinicensis	-	-	I	0.1	I	0.1	I	0.1
Shuteria involucrata var. *glabrata*	I	0.1	-	-	-	-	-	-
Reineckea carnea	I	0.1	I	0.3	II	0.8	III	2.7
Total shrub cover	56.9		65.2		55.7		25.2	
Total herb cover	45.5		40.8		48.3		47.1	

*: Frequency refers to the presence of a species in I: 0<F≤20%; II: 20%<F≤40%; III: 40%<F≤60%; IV: 60%<F≤80%; V: 80%<F≤100%.

**: Dominance refers to relative cover value.

The variation of dominant characteristic and stability of *Quercus spinosa* is one of the outstanding differences among these groups. The dominant characteristic and stability of *Q. spinosa* have declined gradually from group I to group IV, which shows that the role of *Q. spinosa* in the communities has changed from the competitive dominance to the cooperated dominance. In group II, *Q. spinosa* is the specie of dominant characteristic. The variation of species of cooperated dominance is the main difference between groups I and II. The species of cooperated dominance is *Rhododendron rubiginosum* in group I, whereas *Campylotropis capillipes* in group II where the importance of *Q. spinosa* has reduced. And important species includes *Leptodermis forrestii*, *Q. spinosa*, and *C. capillipes*. In the group IV, *L. forrestii* has become the species of prominent dominance, whereas *Q. spinosa* turns into the species of cooperated dominance.

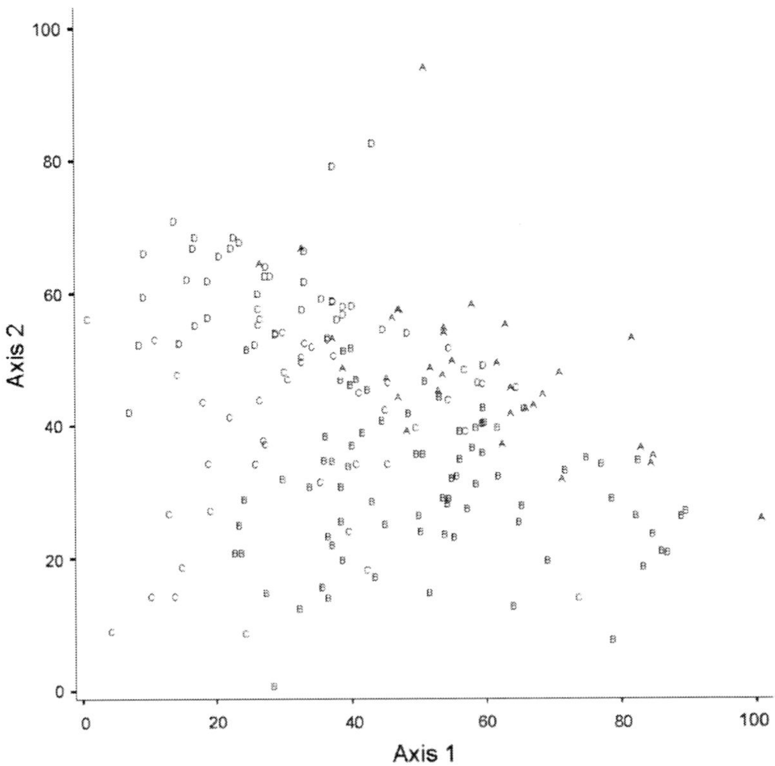

Figure 3.1. Ordination plots for the first two axes of the DCA. The total variance is 1.4491. The first axis explains 18.3% of the variance, the second adds to 12.0%. A: I, B: II, C: III, D: IV.

The DCA ordination of unit data showed that the variation of floristic composition of the four end groups classified by TWINSPAN appeared to be related with the degree of disturbance or the stage of succession and the land coverings (Figure 3.1).

Firstly, group II could be separated from groups III and IV on the ordination axis 1. Compared with the species composition and the total coverings among groups, we can conclude that the gradient of ordination has a tendency of succession. Group III and group IV were located in the early stage of succession series, whereas group I and group II were in the late stage. Secondly, group I and group IV might be separated from group II and group III on the ordination axis 2. The land coverings of group I and group IV are shallow soil, rocky chippings and small stones, which indicates that the gradient of ordination is the land coverings. Based on the land coverings, there are probably two kinds of succession series: one is from group IV to group I where the site is covered with shallow soil and rocky chippings; the other is from group III to group II where the site is covered with deeper soil.

2.3. Characteristics of Individual and Population

Summary of the characteristics of communities is shown in Table 3.2. The density of population and the distribution of *P. armeniacum* vary in the four species communities. *Paphiopedilum armeniacum* was found in the 21 of 32 units for group I, in the 27 of 78 units for group II, in the 15 of 48 units for group III, and in the 12 of 38 for group IV. The density of population was 3.4 individuals per m^2 in group I, 1.2 individuals per m^2 in group II, 1.8 individuals per m^2 in group III and 2.1 individuals per m^2 in group IV. The spatial distribution of population in group I was random, and that of the other three groups was of clustering. It is worthy to be noticed that the clustered degree of distribution of population increased gradually from group II to group IV. The size and grade of distribution were illustrated via the area of leaf per individuals.

The distribution of groups I, II and III has a concave shape, while the distribution of group IV is reverse-J shaped (Figure 3.2). The average area per individual is 0.0043 m^2 in group I, 0.0029 m^2 in group II, 0.0027 m^2 in group III and 0.0023m^2 in group IV.

Table 3.2. Characters of *P. armeniacum* population within each group

Variables	Group I	Group II	Group III	Group IV
Numbers of plot	32	78	48	38
Numbers of individuals	110	94	84	78
Occurrence rate (%)	65.6	34.6	31.3	31.6
Mean individual numbers per plot[a]	3.4(± 0.8)	1.2(± 0.2)	1.8(± 0.5)	2.1(± 0.9)
Variance/mean of numbers of	1.32	1.65	1.87	2.75
Individuals per plot[b]	(1.24)[ns]	(4.04)*	(4.21)*	(7.54)*
Mean leaf area per plant (m^2)	0.0043	0.0029	0.0027	0.0023
c.v. of Mean leaf area per plant (%)	98.5	104.8	121.0	110.1

a: value in parenthesis is standard deviation.

b: value in parenthesis is t-value; * indicate the significant difference ($P<0.05$) between values of Variance/mean and 1; ns indicate no significant difference.

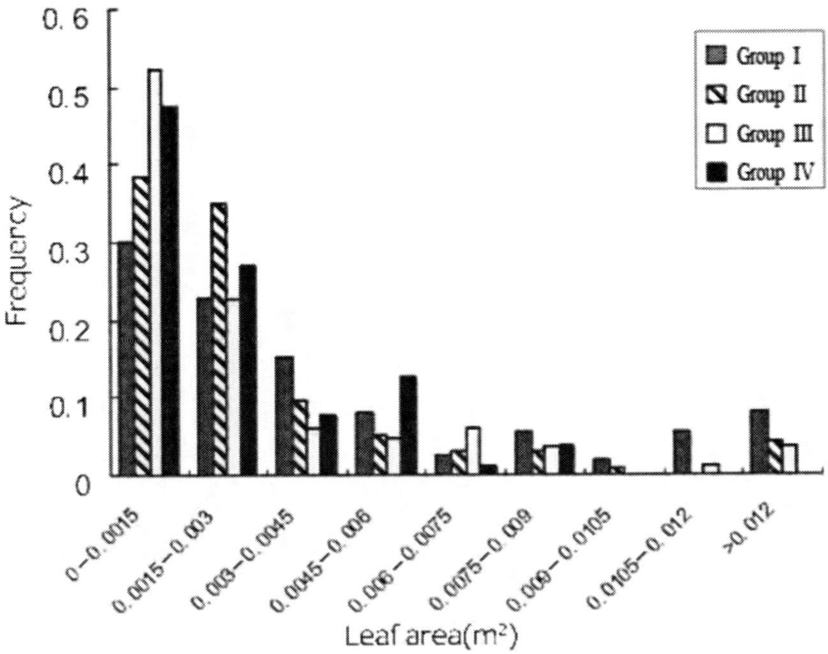

Figure 3.2. The distribution frequency of leaf area in the four end groups of *P. armeniacum*.

3. Discussion

Large scale of *P. armeniacum* can only be found in sites that have experienced long-term disturbance by human beings. The significant difference between the vegetation of present site and absent site of *P. armeniacum* is reflected on the ground coverings, which shows that *P. armeniacum* is a plant of shade-intolerance. And the results of experiment also support this point. Based on the classification and ordination of vegetation plots in Luoshapo Mountain, we can conclude that there are two types of succession series in Luoshapo Mountain: one is from group IV to group I where the site has land coverings of shallow soil, rocky chipping and gravels; the other is from group III to group II where the site has well-developed soil. And the population performance of these two succession series is different. In the former succession series (from group IV to group I), the result has shown that: 1) the density of population is larger at late stage of succession (group I) than at early stage (group IV); 2) the distribution of population at the early stage of succession is agglomerate, whereas at late stage it is randomly distributed; 3) and compared with the size, the average size of individual is smaller and larger ones have disappeared at the early stage during the succession. These results indicated that the population exhibits a tendency from the clonal stage to the expansion stage in former succession series.

However, the result did not show such trend in the latter kind of succession series. The population according to this series has a slight decline in development. The difference of population performance according to different succession series further indicated that *P. armeniacum* is a semi-shaded plant with low competitive abilities in well-developed soil.

Semi-shaded species with low competitive abilities is generally found in early stage of succession when the competition is low. Proper lower competition is probably essential to improve the population performance of such species (Wheeler et al., 1998; Wotavova et al., 2004). The result of our study also confirms this point of view. Disafforestation and grazing have slowed down or impeded the growth of other species of highly competitive abilities and brought up the habitat of *P. armeniacum* on Luoshapo Mountain. As for *P. armeniacum*, habitat with poor condition of disafforestation is more appropriate for its growth. Disafforestation and grazing might entirely destruct the habitat and it is propitious to maintain the survival of *P. armeniacum* population by enhancing the natural succession (especially that of group IV to group I) in such region. Therefore, it is necessary to manage vegetation of various densities. To a large extent, the change of land use is one cause for the change of quality in habitat, which recently attracts more and more attention of studies. Due to various degrees in the use of land, the landscape presents a plot-embedded shape at different stages of secondary succession, which provides a chance for maintaining the high ecological diversity. However, the traditional usage of land has disappeared fastly, and the implementation of law and the change of social economy have extended the natural succession. For example, thanks to the National Forest Project, more and more land in the area of our study has turned into forest and the significant change is that trees grow taller and the overcast rate of forest has increased in most mountains. So the landscape becomes uniform. Thereby, the conservation for ecological diversity of traditional usage of agricultural land becomes valueless, which could probably increase the risk of relative species in habitat and at early stage of succession. *Paphiopedilum armeniacum* is one of outstanding examples.

Because of the special pattern of population distribution of *P. armeniacum*, the relative centralization of flowers increases the rate of pollination, but there are special requirements for its habitat, so there are few sites in the distributive region suited for its growth. This special way of growth has limited the expansion of its population, which may be the reason that *P. armeniacum* displays narrow distribution while it is provided with the ability of expansion. *Paphiopedilum armeniacum* has effective strategy of sexual reproduction, so that a considerable amount of flowering plants can fruit and there are numerous seeds in each fruit; although only a few seeds could grow into seedlings, a few gains from mass sowing is enough to maintain the stability and the growth of population. This measure takes asexual reproduction into consideration when it devotes the accumulative resources from vegetative growth to the sexual reproduction. This effective strategy of breeding enables the population to maintain its stability and even keep its continuance and development in vegetative environments of shallow poor soil and sunshine penetration.

Population performance of plants is the outcome of interreaction between the individual viability and the outside environment (Crawley, 1986). Population performance in different habitats not only reflects the actual condition of population, but also exhibits the relationship of conflict between population and environment (Manuel and Molles, 2002). The key of its success in the conflict with environment is attributed to its successful breeding behaviors. Its sexual reproduction generates progeny many times over a long period, so that it becomes the foundation of development of population, which indicates that breeding strategy is the main factor that influences the change of population in quantity (increase or descrease), thus affects the mating mode.

So it can make response to the habitat that influences population via the change of mating mode.

REFERENCES

Blomqvist, M. M., Vos, P., Klinkhamer, P. G. L., et al. (2003). Declining plant species richness of grassland ditch banks – a problem of colonisation or extinction? *Biological Conservation*, *109*, 391–406.

Brower, J. E., Zar, J. H., von Ende, C. N. (1998). *Field and laboratory methods for general ecology*, 4th edition. Boston: Mcgraw-Hill.

Ceballos, G. and Ehrlich, P. R., (2002). Mammal population losses and the extinction crisis. *Science*, *296*, 904–907.

Chen, X. Y. (2000). Effects of plant density and age on the mating system of *Kandelia candel* Druce (Rhizophoraceae), a viviparous mangrove species. *Hydrobiologia*, *432*, 189–193.

Crawley, M. J. (1986). *Plant Ecology*. London: Blackwell Scientific Publications.

Fahrig, L. (1997). Relative effects of habitat loss and fragmentation on population extinction. *Journal of Wildlife Management*, *61*, 603–610.

Farris, M. A. and Mitton, J. B. (1984). Population density, outcrossing rate, and heterozygote superiority in *Ponderosa pine. Evolution*, *38*, 1151–1154.

Honnay, O., Bossuyt, B., Verheyen, K., et al. (2002). Ecological perspectives for the restoration of plant communities in European temperate forests. *Biodiversity and Conservation*, *11*, 213–242.

Karron, J. D., Thumser, N. N., Tucker, R., et al. (1995). The influence of population-density on outcrossing rates in *Mimulus Ringens*. *Heredity*, *75*, 175–180.

Lin, Z. and Xie, Z. (2005). Does habitat restoration cause species extinction? *Biological Conservation*, *123*, 349–354.

Manuel, C. and Molles, J. (2002). *Ecology, Concept and Applications*, 2nd edition. New York: McGraw-Hill Companies.

McCune, B. and Mefford, M. J. (1999). PC-ORD Software for Multivariate Analysis of Ecological Data. Version 4.0 MjM Software, Gleneden Beach, Oregon, USA.

Murawski, D. A. and Hamrick, J. L. (1991). The effect of the density of flowering individuals on the mating systems of nine tropical tree species. *Heredity*, *67*, 167–174.

Ne'eman, G. (2003). To be or not to be—the effect of nature conservation management on flowering of *Paeonia mascula* (L.) Miller in Israel. *Biological Conservation, 109*,103–109.

Watters, J. V., Lema, S. C., Nevitt, G. A. (2003). Phenotype management: a new approach to habitat restoration. *Biological Conservation, 112*, 435–445.

Webb, J. K. and Shine, R. (2000). Paving the way for habitat restoration: can artificial rocks restore degraded habitat of endangered reptiles? *Biological Conservation, 92*, 93–99.

Verheyen, K. and Hermy, M., (2004). Recruitment and growth of herb-layer species with different colonizing capacities in ancient and recent forests. *Journal of Vegetation Science, 15*, 125–134.

Wheeler, B. D., Lambley, P. W., Geeson, J. (1998). *Liparis loeselii* (L.) Rich in eastern England: constraints on distribution and population development. *Botanical Journal of the Linnean Society, 126*, 141–158.

Wotavova, K., Balounova, Z., Kindlmann, P. (2004). Factors affecting persistence of terrestrial orchids in wet meadows and implications for their conservation in a changing agricultural landscape. *Biological Conservation, 118*, 271–279.

Chapter 4

Asexual Reproduction of Paphiopedilum Armeniacum

Abstract

Paphiopedilum armeniacum has two modes of asexual reproduction—by tillering and by producing rhizomes. Tillering of asexual reproduction is of strong vitality and is endowed with an emergency mechanism to cope with adversity by producing numerous rhizomes. The purpose of clonal reproduction is to extend the lifespan of genets, to implement the sexual reproduction and to increase the proportion of sexual mating. The rhizome reproduction in *P. armeniacum* is also an adaptation to the condition of defoliation or to the condition covered with soil in order to escape from death. *Paphiopedilum armeniacum* still keeps the original reproductive pattern in *ex situ* habitat with an obviously different condition from the original one. Clonal genets also keep the patterns of growing, surviving and breeding, which are adaptive to the uncertain conditions after returning to the original habitat. All further indicate that its reproductive mode has evolved from ecological adaptation to genetic adaptation.

Asexual reproduction is common in plants that can completely free plants from the dependance on pollinating agents, and its main way is vegetative reproduction. Vegetative reproduction doesn't need to go through a status of single cell called 'bottleneck' (the combination of ovum and sperm), and it is more often treated as a way of growth rather than a way of reproduction because vegetative reproduction is not created by information producing from code in a cell (Harper, 1977). However, clonal reproduction has endowed clonal plants with many special characteristics, such as activity, endurance, abilities of spatio-temporal extension and multiplication, the physiological integration and risk apportion among ramets on the same genet (Cook, 1983; Eriksson and Jerling, 1990) and the ability of rapidly absorbing limited resource and saving it for future use (Chapin et al., 1990). What should not be neglected is that the relation between this vegetative reproduction and sexual reproduction and its effect on the ensurance of survival that can cope with unfavorable habitat. It's because that, in nature, a plant can produce individuals with potential self-help viability through the way of asexual reproduction (Mogie and Hutchings, 1990; de Kroon and van Groenendael, 1997), and the clonal growth of plant can be represented by many ways, such as producing ramets from creeping stem on the ground or from underground rhizome. The level component of clonal plant has the potential of self-help viability that can separate from mother plant, so that the vertical component of non-clonal plant is always in conformity with mother plant physically and physiologically.

As a result of this characteristic, the level component of clonal plant comes into being vegetative offspring (ramet) and increases the quantity of individuals in population; only according to this, it is namely breeding (Bazzaz and Grace, 1997).

The clonality is a very original character in terms of evolution, widely existing in various classified populations (Mogie and Hutchings, 1990). The clonality of angiosperm is probable hypo-created, that is, evolves from non-clonal ancestor.

Clonality and the ways of clonal growth that plant uses to achieve clonality have a series of meanings in ecology and evolution (Silander, 1979; Stuefer et al., 2002).

Though many plants have both sexual reproduction and clonal growth, and the balance between these two breeding ways is great in different species and among different populations of a same species, few studies have worked over the effect on sexual reproduction caused by the influences of the clonal ways, the evolution, the inheritance and ecological aspect of clonal plants (Charpentier et al., 2000; Eckert, 2000, 2002; Dorken and Eckert, 2001).

Thus, studies on the asexual growth, survival and the inner relationship between clonal reproduction and sexual reproduction of *P. armeniacum* are to reveal the evolutionary meaning of adaptation to habitat from its asexual reproduction.

1. METHODS

1.1. Observation on Way of Asexual Production in Original Habitat

During the observation of flowering status, the asexual reproduction of *P. armeniacum* was also simultaneously observed in each site, and the pattern of breeding, quantity, and process of growth of clonal ramets on each genet were recorded.

1.2. Validation of Clonal Way

120 genets selected randomly from the original habitat in Yunnan were transported to Shenzhen and planted (divided into 4 groups, 30 genets each group, n = 30) respectively in the artificial spare woods (with canopy density of 0.4 – 0.6 and good drainage) and in the large dense forest (canopy density of nearly 1 and good drainage). Two groups in the form of 30 cm × 30 cm were planted in the artificial spare woods: embedding the whole root but exposing the sheath basal, and then one of this two groups was covered with humus up to the central region of leaves (deep-buried group, n = 30), and the other group without any treatment (shallow-planted group, n = 30). The other two groups were planted in the large dense forest in the same way as in the spare woods. The status of growth and breeding of *P. armeniacum* in each treatment were then observed.

1.3. Tests of *Ex Situ* and Reintroduction Reproduction

In order to check the growing state of *P. armeniacum* in the *ex situ* site and in the original habitat after reintroduction, 10 genets were collected (during the flowering stage but not during fruiting) from each sampled site (No. 1 – 8) in May 2000, and were transferred as bare-rooted genets to the *ex situ* site, the National Orchid Conservation Center in Shenzhen. The genets from each sampled site were divided into 2 groups (5 genets for each group) for the test of asexual reproduction. One group was cultured in pots (n = 8, 40 genets in total), and the other was directly planted in sandy soil (n = 8, 40 genets in total). The number of reproduced ramets was counted in May of

2001 and 2002. In April 2003, the added ramets from pot reproduction were transported back to the original habitat (n = 8), and 20 ramets (genet of single bud) from each sampled site were separately planted, watered once after planted. Their status of growing and flowering were monitored and recorded in May and October of 2004 and 2005.

2. Result

2.1. Characteristics of Asexual (Clonal) Reproduction

There are two modes of asexual (clonal) reproduction of *P. armeniacum*: by tillering new ramet from the base of ramet after flowering (Figure 4.1) and by producing rhizomes from the base of ramet, and then they grow into new genets after breaking the ground.

Figure 4.1. Tillering reproduction of *P. armeniacum* is an usual mode of asexual reproduction when it is not covered. Tillering buds generate from the base of ramet after flowering to form a new ramet that has close resource integration with its mother plant.

Tillering reproduction is the usual mode of asexual reproduction used by genet of normal growth: a seedling or ramet will bloom once it has 3 – 6 (or 8) leaves, and ramet (genet) after flowering will generate tillering buds from the sheath base. The tillering bud will germinate alabastrum once it grows 3 – 6 leaves. One genet may grow 2 or 3 tillering buds at the same time, and ramets of fasciculate genet with several buds can all germinate tillering buds after flowering during that time. And ramets without sexual phase (post-flowering) usually do not have tillering reproduction.

Paphiopedilum armeniacum has very strong capability of rhizomic reproduction when covered. After covered by shatter and earth, the rhizome will germinate from the base of genet. And the rhizome has the apical dominance and gradually develops and elongates so that it can break through the covering and then grows into a new individual. The length of rhizome is from 3 – 5cm to 26 – 30cm, which depends on the depth of being buried. One genet can produce several rhizomes, and each of them can grow into a new plant. The rhizome will grow sheath and leaves after breaking through the covering and produce new roots at the base of sheath, and then a new individual is shaped. The rhizome will rot and wither after the apex has grown into new ramet, and the new ramet will detach from the mother plant and become a new, independent individual, meanwhile the mother genet will die by being covered. Genets seriously covered will die because the rhizomes cannot break through the covering. Genets exposed can only grow tillering buds but not rhizomes. Genets that are pulled out or washed out by rain and then placed back on the ground surface once more will produce rhizome in wet and shady area. Individual from rhizomic reproduction will mature and flower in 2 years (Figure 4.2).

2.2. Validation of Way of Clonal Reproduction

After one month in the artificial spare woods, the living genets of shallow-planted group (SPG) all could produce tillering buds and those of deep-buried group (DBG) all could produce rhizomes. The reproduction modes are shown in Table 4.1. There is significant difference in quantity of tillering buds between SPG and DBG ($t = 2.971$, $d.f. = 58$, $P < 0.001$), so is the quantity of rhizomes ($t = 3.082$, $d.f. = 58$, $P < 0.001$). The results show that the production of rhizome is highly correlated with the covered degree of genet, which is consistent with the result of field research.

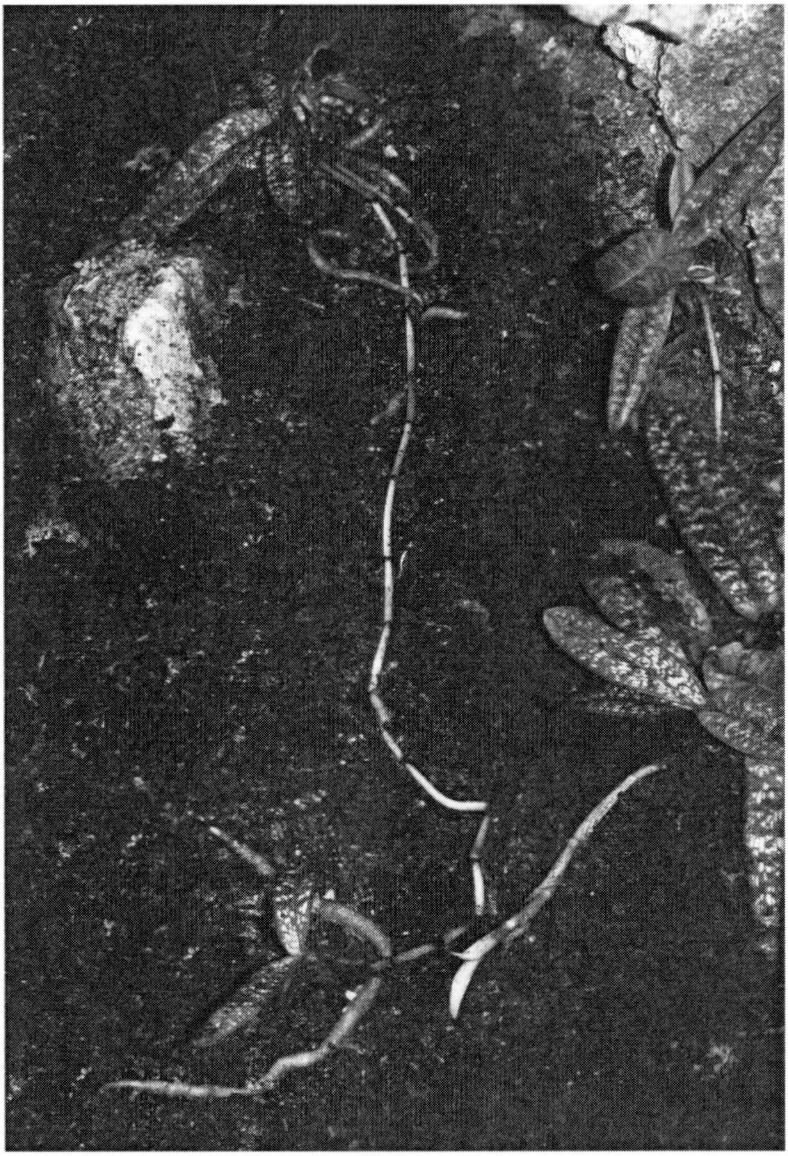

Figure 4.2. Rhizomic reproduction of *P. armeniacum*. It is an emergency way of reproduction when it is covered. It can produce several rhizomes, of which some are 40 cm long, and then it breaks through the covering to form new ramet, so that it can escape from harmful habitat.

Table 4.1. Reproductive Mode (RM) of *P. armeniacum* in SPG and DBG in spare forest

Group	RM	Sample no.																														$\overline{x}$ ±SD
		1	2	3	4	5	6	7	8	9	10	11	12	13	14	15	16	17	18	19	20	21	22	23	24	25	26	27	28	29	30	
SPG	Tiller	1	2	3	2	1	3	3	2	1	3	2	2	2	1	2	3	2	1	3	2	2	2	2	1	1	1	2	3	3	3	2.03±0.76
	Rhizome	0	0	0	0	0	0	0	0	0	0	0	0	0	0	0	0	0	0	0	0	0	0	0	0	0	0	0	0	0	0	0
DBG	Tiller	0	0	0	0	0	0	0	0	0	0	0	0	0	0	0	0	0	0	0	0	0	0	0	0	0	0	0	0	0	0	0
	Rhizome	2	3	4	5	2	3	4	3	4	3	2	3	3	1	2	4	3	3	2	2	3	3	2	2	3	3	4	2	1	3	2.80±0.92

Table 4.2. Statistic of *P. armeniacum* viable rate cultivated in DF and SF

Sample no.	1	2	3	4	5	6	7	8	9	10	11	12	13	14	15	16	17	18	19	20	21	22	23	24	25	26	27	28	29	30	$\overline{x}$ ±SD (%)
SF	100	0	100	100	0	100	100	100	100	100	100	100	100	100	100	100	100	100	100	100	100	100	100	100	100	100	100	100	100	100	93.33±25.37
DF	0	0	0	0	0	0	0	0	0	0	0	0	0	0	0	0	0	0	0	0	0	0	0	0	0	0	0	0	0	0	0

Genets in dense forest (DF) or in spare forest (SF) all could produce new buds or rhizomes in a month, but the buds or rhizomes of dense forest were etiolated, weak and would gradually wither. The survival rates are shown in Table 4.2. There is significant difference in the survival rate between genets of spare forest and that of dense forest (n = 30) (t =3.551, $d.f.$ = 58, $P<0.001$). The result indicates that *P. armeniacum* can grow well in spare forest but cannot grow in dense forest of large trees, which was consistent with the result of field observation.

2.3. Reproduction of *Ex Situ* and Reintroduction

2.3.1. Ex situ Reproduction

Paphiopedilum armeniacum in pot could produce many rhizomes or tillering buds in that very year, and will flower in 2 years. It could fruit and produce seeds after artificial cross-pollination or self-pollination. Ramets would germinate tillering buds after flowering. *Paphiopedilum armeniacum* planted in soil had the same results as the potted *P. armeniacum*. The breeding status are shown in Tables 4.3 and 4.4. The results show that *P. armeniacum*, either planted in pots or in soil, are able to carry out effective asexual reproduction.

Table 4.3. Reproductive state of cultivated *P. armeniacum* in pot of eight sites

Site no.	Quantity of genet	Quantity of cloned ramet in 2000–2001		Quantity of cloned ramet in 2002		Σ	Growth rate (%)
		Tiller	Rhizome	Tiller	Rhizome		
01	5	0	15	2	0	22	340
02	5	1	8	1	1	16	220
03	5	2	12	3	0	22	340
04	5	3	4	1	2	15	200
05	5	0	14	2	0	21	320
06	5	0	18	3	0	26	420
07	5	2	9	2	1	19	280
08	5	1	8	2	0	16	220
Σ	40	9	88	16	4	157	2340
$\overline{x}$	5	1.13	11	2	0.5	19.63	292.50
SD	0	1.13	4.57	0.76	0.76	3.81	76.30

Table 4.4. Reproductive state of cultivated *P. armeniacum* in forest of eight sites

Site no.	Quantity of genet	Quantity of cloned ramet		Quantity of cloned ramet		Σ	Growth rate (%)
		Tiller	Rhizome	Tiller	Rhizome		
01	5	0	15	2	1	23	360
02	5	1	8	1	2	17	240
03	5	2	12	2	0	21	320
04	5	3	4	3	0	15	200
05	5	0	14	2	0	21	320
06	5	0	18	1	1	25	400
07	5	2	9	2	1	19	280
08	5	1	8	1	1	16	220
Σ	40	9	88	14	6	157	2340
$\overline{x}$	5	1.13	11	1.75	0.75	19.63	292.50
SD	0	1.13	4.57	0.71	0.71	3.50	70.05

2.3.2. Reproduction of Reintroduction

Based on the statistics in 2004, the survival rate of genets after reintroduction was 38.13 ± 13.91% (n = 8). The living genets could normally grow and flower in the open and their flowering rate was 14.38 ± 9.16% (n = 8).

Eristalis tenax, *Ceratina morawitzi* and *Lasioglossum pronotale* were found to visit flowers of *P. armeniacum* during the flowering period. The flowers produced fruits and the fruiting rate was 35.83 ± 30.31 % (n = 8) of the flowering ramets. And the blooming rate of 2005 was 26.25 ± 9.60% (n=8).

The status of growth and flowering of *P. armeniacum* that were reintroduced to the native habitat are shown in Table 4.5. The results have shown that there was no significant difference between the fruiting rate of ramets reintroduced to the original habitat and natural fruiting rate in wild (t =0.1878, $d.f.$ = 14, P = 0.8537).

Table 4.5. Growing and flowering states of *P. armeniacum* replanted in original habitat

Site no.	Quantity of replanted genet	2004				2005			
		Quantity of viable genet	Quantity of cloned ramet	Quantity of flowering ramet	Quantity of fruiting ramet	Quantity of cloned ramet	Quantity of flowering ramet	Quantity of fruiting ramet	Fruit set rate (%)
01	20	11	31	5	1	36	8	3	37.50
02	20	7	12	1	0	15	3	1	33.33
03	20	9	20	3	1	22	7	3	42.86
04	20	4	28	4	2	28	7	2	28.57
05	20	6	11	2	1	14	3	1	33.33
06	20	8	16	1	0	18	5	2	40.00
07	20	12	20	6	2	23	6	1	16.67
08	20	4	7	1	1	10	3	1	33.33
Σ	160	61	145	23	8	166	42	14	265.59
$\overline{x}$	20	7.63	18.13	2.88	1	20.75	5.25	1.75	33.20
SD	0	2.97	8.34	1.96	0.76	8.40	2.05	0.89	8.04

3. Discussions

Paphiopedilum armeniacum has both sexual reproduction and vegetative reproduction; offspring of vegetative reproduction (ramets) could distribute either sparsely (rhizomic reproduction) or densely (tillering reproduction), and they have some physical connection and physiological conformity with their mother genets. And this connection and conformity keep tightly and last longer in ramets of tillering reproduction but last shorter in ramets of rhizomic reproduction. In emergent condition, it selectively sets the ramets in place of good condition in form of rhizomes, and the cost is the death of mother genet. The resource expense of rhizomic reproduction is greater than that of tillering reproduction, so rhizomic reproduction can only be used when it has no other choice. As for genets, the clonal growth can surely help genets to escape from aging due to its continuous splits of asexual ramets; and as for ramets, the outcome of resource accumulated from its vegetative growth is to maximally achieve the effect of sexual reproduction and produce new ramets after sexual reproduction, but self-aging is the result of resource expense.

Deep-buried *P. armeniacum* can only grow rhizomes and those plants of shallow-planted only produce tillering buds, which is a survival reaction of *P.*

armeniacum to the environment. In wild, genets of *P. armeniacum* are often covered by shatters and loose earth, or swept away entirely by rain, so they can only survive after breaking through the superstratum or re-entering into the earth, or they will die. Therefore, *P. armeniacum* produces rhizomes to multiple new individuals, which is another mode of rhizomic reproduction, in fact it is also a kind of survival assurance evolved from the adaptation to environment, by which it overcomes unfavorable environment and even escapes from death.

The fruited rate of flowering ramets of *P. armeniacum* is about one-third, not very high, due to its imitation of the flower of associated plants to attract pollinating insects. Such pollinating skill is much more evolutionary, but it is easily intermitted by environmental factors, however, the quantity of seeds in one capsule can reach 50,000, which indicates that the sexual reproduction is still effective and the clonal growth must occur after sexual reproduction. Therefore, it can be concluded that the clonal growth of *P. armeniacum* is totally for the benefit of sexual reproduction and it will not result in the declination or loss of sexual reproduction.

Mating system is the pattern that an organism uses to pass its germs from one generation to the next generation via sexual reproduction, including all characteristics of controlling the combination of gametes to bring up zygote (Barrett and Eckert, 1990). There are three kinds of mating modes in orchids, including selfing, crossing and the mixed mating (Darwin, 1890). These mating modes are controlled by genetic substance and they also have close relation with the environmental conditions. For instance, *Holcoglossum amesianum* is a plant that reserves the floral structure for cross-pollination but it can possess self-pollination and self-fertilization at the same time, and it can automatically carry out self-pollination in conditions of drought and lacking pollinating insects (Liu et al., 2006). As *P. armeniacum* will be confronted with unfavourable climatic conditions unpredictably in the habitat during the process of flowering and pollination, and in order to avoid the risk of failure in breeding caused by these conditions and to safeguard the extension of its life, this species has evolved another mechanism to achieve successful breeding via its asexual reproduction, in which the ramets can produce asexual buds to enlarge the quantity of community, no matter whether there is a success in flowering or seed producing or not (Zhang, 2004), accordingly, to increase the quantity of sexual reproduction in the community finally. Therefore, we consider that the assurance of sexual reproduction is the evolutionary motivation of asexual reproduction; at least it's the breeding ways of *P. armeniacum*.

No organism can make every thing work well at one time. *Paphiopedilum armeniacum* carries out asexual reproduction and sexual reproduction annually and has very good effect from this kind of resource input which may bring up no cost in the survival rate at the later period of its life cycle (Zhang, 2004). *Paphiopedilum armeniacum* has found out a best solution for effective allocation of resource among the conflicting demands of limited energy for its growth, survival and breeding, that is, to further sexual reproduction via the reinforcement of asexual reproduction. If asexual reproduction has been dropped out, it will take very long time for *P. armeniacum* to bring about sexual reproduction one more time; after all, it takes much more time from seed germinating to flowering than the time an asexual ramet required for flowering. And there are risks in the process of spreading seeds to finding suitable habitat, as well as in the process of germination, growth and florescence in the suitable habitat. Perennial plant is much worthwhile than annual plant in the light of avoiding risks of breeding, either temporally or spatio-temporally. The reproductive strategy of *P. armeniacum* shows that ecological factors may have influence on the variation of sexual reproduction of this species and facilitate this plant in heterogeneous environment that can grow in conditions of strong light, large differences in temperature and seasonal drought (Tang et al., 2007), and of which breeding mode has balanced the advantages of sexual reproduction and asexual reproduction to establish foundations for the genetic evolution (Klekowski, 1988).

REFERENCES

Barrett, S. C. H. and Eckert, C. G. (1990). Variation and evolution of mating systems in seed plants. In: Kawanos, ed. *Biological Approaches and Evolutionary Trends in Plants*. Oxford: Academic Press, 229–254.

Bazzaz. F. A. and Grace J. B. (1997). *Plant Rescource Allocation*. San Diego: Academic Press.

Chapin, F. S., Schulze, E. D., Mooney, H. A. (1990). The ecology and economics of storage in plants. *Annual Review of Ecology and Systematics*, *21*, 423–447.

Charpentier, A., Grillas, P., Thompson, J. D. (2000). The effects of population size limitation on fecundity in mosaic population of the clonal macrophyte *Scirpus maritimus* (Cyperaceae). *American Journal of Botany*, *87*, 502–527.

Cook, R. E. (1983). Clonal plant-populations. *American Scientist*, *71*, 224–253.

Darwin, C. (1890). *The Various Contrivances by which Orchids are Fertilised by Insect*, 2nd edition. London: John Murray.

de Kroon, H. and van Groenendael, J. (1997). *The Ecology and Evolution of Clonal Plants*. Leiden: Backhuys Publishers.

Dorken, M. E. and Eckert, C. G. (2001). Severely reduced sexual reproduction in northern populations of a clonal plant, *Decondon verticillatus* (Lythraceae). *Journal of Ecology*, *89*, 339–250.

Eckert, C. G. (2000). Contributions of autogamy and geitonogamy to self-fertilization in a mass-flowering, clonal plant. *Ecology*, *81*, 532–542.

Eckert, C. G. (2002). The loss of sex in clonal plants. *Evolutionary Ecology*, *15*, 501–520.

Eriksson, O. and Jerling, L. (1990). Hierarchical selection and risk spreading in clonal plants. In: van Groenendael J, de Kroon H., ed. *Clonal Growth in Plants: Regulation and Function*. The Hague: SPB Academic Publishing.

Harper, J. L. (1977). *Population Biology of Plants*. London: Academic Press.

Klekowski, E. J. J. (1988). *Mutation, Developmental Selection and Plant Evolution*. New York: Columbia University Press.

Liu, Z. J., Liu, K. W., Chen, L. J., et al. (2006). Conservation ecology of endangered species *Paphiopedilum armeniacum* (Orchidaceae). *Acta Ecologica Sinica*, *9*, 2791–2800.

Mogie, M. and Hutchings M. J. (1990). Phylogeny, ontogeny and clonal growth in vascular plants. In: van Groenendael J. and de Droon H., ed. *Clonal Growth in Plants*. The Hague: SPB Academic.

Silander, J. A. (1979). Microevolution and clone structure in *Spartina patens*. *Science*, *203*, 658–660.

Stuefer, J. F., Erschbamer, B., Huber, H., et al. (2002). The ecology and evolutionary biology of clonal plants: an introduction to the proceedings of Clone-2000. *Evolutionary Ecology*, *15*, 223–230.

Tang, Z. H., Yang, L., Liang, S. N., et al. (2007). Effects of different water conditions on life cycle forms and physiological metabolisms of *Catharanthus roseus*. *Acta Ecologica Sinica*, *27* (7), 2742–2747.

Zhang, D. Y., ed. (2004). *Ecology of Evolution and Reproduction on Plant Life History*. Beijing: Science Press, 1–421.

Chapter 5

SEXUAL REPRODUCTION OF PAPHIOPEDILUM ARMENIACUM

ABSTRACT

Paphiopedilum armeniacum is able to produce many seeds through sexual reproduction and the seeds can grow into new genets effectively. It deceives various pollinators among populations to achieve selfing, crossing, or even mixed mating in flowers by mimicking flowers of associated plants. In conditions of unsaturated pollinators or external pollen, self-pollination completed by pollinators could increase seeds production. The high inbreeding depression in the self-pollinating estimate of this study showed that, as the pollinating environment turned into an essential factor of reproductive assurance for wild species, the increase of the selfing rate could increase the effect of reproductive assurance. We demonstrated how the visiting behavior of pollinating insects in a changeable pollinating environment influenced the reproductive assurance. This demonstration indicates that flowers with a strict cross-pollinating structure also have a selfing mechanism and a changeable pollinating environment is propitious to the selfing and the mixed mating, which associates the reproductive assurance with the selfing rate of flowers of non-autonomous self-pollination via the mechanism of mixed mating to display the diversity of mixed mating evolution. The mix-breeding mode of *P. armeniacum* undounbtedly reflects the success in evolution of flowering plants in terms of the diversity of breeding strategies, and it also gives an insight into why the vast majority of plants have self-compatibility, or most plants employ reproductive strategies of selfing or mixed mating, although crossing is of superiority.

The phenomenon that a plant produces descendants via selfing and crossing brings a challenging question to biologists in evolution (Goodwillie et al., 2005). There is selective advantage in those individuals that can carry out self-pollination but haven't yet completed cross-pollination of cross-pollinating plants, which is achieved by animal agents when pollinators are not reliable (Lloyd, 1980). Therefore, the reproductive assurance is one of the essential factors for the evolution of self-pollination (Holsinger, 1996). During the evolution of mating system promotion, the effect of reproductive assurance via self-pollination is greater than that via inbreeding depression, so the mating system of mixed selfing and crossing will be kept down (Kalisz et al., 2004; Kalisz et al., 1999). If the opportunity of crossing is reduced, the breeding of a plant should rely even more on selfing (Jain, 1976; Lloyd, 1992; Schoen and Brown, 1991). Plant has a countermeasure of reproduction that can react to the changes of environment (Holsinger, 1991; Liu et al., 2007), for example, populations of flowering plants which lack mating partners or pollinators and populations with uncertainly receivable amounts of cross-pollen in a year or among years can evolve autonomous within-flower self-pollination having no pollinating agents (Lloyd, 1992; Schoen and Brown, 1991; Kalisz et al., 2004; Liu et al., 2006; Fausto et al., 2001; Barrett, 2003). Theoretically, such a condition can also bring up within-flower self-pollination having pollinating agent. If so, its evolution is also a trade-off between the advantage of selfing and the potential cost. Under the condition that the summation of various costs (including seed discounting, pollen discounting) (Lloyd, 1992; Schoen et al., 1996) and inbreeding depression (Jarne and Charlesworth, 1993) must not balance out the fitness that gained from self-pollination, even though self-pollination can provide reproductive assurance (Schoen and Brown, 1991; Kalisz et al., 2004; Herlihy and Eckert, 2002). And those costs are changed along with the following two factors: the received duration (Lloyd, 1992) of external pollen compared with self-pollination and the validity of pollinators (Lloyd and Schoen, 1992; Arroyo et al., 1981). If there are cross-pollinators and the behaviour of self-pollinators takes place after all possible opportunity of crossing (delayed selfing), selfing would not cause any pollen dicounting or seed discounting (Lloyd, 1992; Schoen and Brown, 1991). Though the inbreeding depression (δ) is very high, the survival of descendants from delayed selfing can always provide reproductive assurance (Schoen and Brown, 1991). If selfing and the receipt of external pollen occur at the same time (competing selfing) and when there is sufficient external pollen, seed discounting, pollen discounting and inbreeding depression will all exist, which indicates that selfing is harmful (Lloyd, 1992; Fausto et al., 2001); on the other hand, when there is not sufficient external pollen to fertilize all ovules, selfing could provide reproductive assurance more or less (Lloyd, 1992; Herlihy and Eckert, 2002). When there is a shortage of cross-pollinator, selfing completed by self-pollinators does not bring up seed discounting or pollen discounting (Holsinger, 1991; Lloyd, 1992; Schoen and Brown, 1991; Kalisz et al., 2004); and when there is neither cross-pollinators nor self-pollinators, the plant would lose all costs of breeding. Based on our field study, the results of detection on pollinating mechanism and reproductive system of *P. armeniacum* show that this species can successfully imitate flowers of associated plant to deceive various cross-pollinators and self-pollinators to visiting its flowers, which results in cross-pollination, self-pollination, and both self-pollination and cross-pollination (within flowers) among populations so that it can bring out a mixed mating system. There is spatial isolation between male and female functions of *P. armeniacum* (Schoen and Brown, 1991), so autonomous selfing will not happen, but its structure of cross-pollination (Holsinger, 1996) can assure cross-pollination and achieve self-pollination as well. The function of cross-pollination accompanies with the whole flowering period but the function of self-pollination just starts in the delayed period (Arroyo et al., 1981), and the phenotype of both cross-fertilization and self-fertilization indicates that it has various successful reproductive strategies, which reflects that the self-pollination is evolved from an original species of cross-pollination on a special level within genus or species (Vasek and Weng, 1988). This study investigated the visiting behaviour of pollinators in four reproductive seasons and the result shows that *P. armeniacum* has to try its best to use self-pollinators to assure successful reproduction in the condition of insufficient or unreliable cross-pollinators, thus it evolves a mixed mating system. This is because the natural selection is beneficial to those individuals with strategy of reproductive assurance and less depending on special pollinators (Totland and Sottocornola, 2001).

1. METHODS

1.1. Observation Site

The field observation and experiments were implemented in 10 populations of *P. armeniacum* (8 in Laowo village of Lushui Country, 1 in Yangliu of Baoshan City and 1 in Minjian of Yunlong Country) at an altitude of 1600 – 2250 m on Nushan Mountain in Yunnan Province during 2003 – 2007.

1.2. Observation on Biological Characteristics

1.2.1. Flowering Phenology and Characteristics of Floral Configuration

The flowering phenology of all plants of *P. armeniacum* in the observed natural community was observed and recorded during the 5 flowering seasons from 2003 to 2007, including the beginning of flowering period, the full flowering stage, and the final flowering stage respectively. 20 plants with unopened flowers were randomly selected each year and marked for observation to record the following suite of variables: the number of flowers (buds) on each inflorescence, the length of inflorescence, and the flowering characteristics of single flower which includes the opening time of each flower, the fading time (of corolla), the splitting time of anther sac and its change, and the morphological changes of both labellum and corolla.

1.2.2. Observation on the Morphological Structure of Pollen Grains and Stigma and the Relative Position of Anther and Stigma

Stamens and stigmas were cut down from freshly opened flowers, and their morphologies and the relative position of anther (pollen sac) and stigma were examined directly under stereoscope (Guiguang XTL-500, China) and then were photographed.

The stigma and the dehiscent anther or anthers with viscid envelopes were fixed with 4% glutaric dialdehyde, and then were scoured off the fixed liquid with 0.2 mol/L phosphate buffer solution, dehydrated with ethanol and isoamyl acetate, and dried at CO_2 critical points, and then pasted and mounted with gold. The stigma surface and the morphological characteristics of pollen grains were observed under the scanning electron microscope (Philips XL-30 ESEM, Netherland) and photographed.

1.3. Detection on Characteristics of Flowering Phenology

1.3.1. Tests of Quantities of Pollen Grains and Ovary and the Pollen/Ovule (P/O) Ratio

Twenty plants in buds (with un-split anther) were randomly selected and cultivated in the laboratory for backup use. Anthers from each bud were carefully picked down and all pollen grains were transferred into calibrated tubes and filled with 0.5 – 1 ml distilled water and then vibrated for 60 seconds in a shaker. 1μl of pollen solution was transferred onto the slide and observed under microscope (160×) to count the quantity of pollen grains (Wuzhou, XTJ-4400, China). The experiment of each anther was repeated 10 times. At the same time, ovary of each flower was placed on the slide and dissected under an anatomical lens (40×) for observation and the quantity of ovules was counted. The quantity of pollen grains divided by the quantity of ovules is the pollen/ovule (P/O) ratio of each flower.

1.3.2. Tests of the Receptivity of Stigma and the Viability of Pollen

Fresh flowers were randomly selected after opening, respectively in 1 day, 5 days, 10 days, 15 days, 20 days, 25 days, 30 days, and 35 days for backup use. The viability of pollen was identified by the MTT method (Rodriguez-Riaño and Dafni, 2000): the fresh pollen were smeared on a slide and compared with dead pollen (by boiling). A drop of MTT solution was added into the pollen specimen and mixed fully, and another drop of MTT solution was added after air-dry. The dry specimen was observed under microscope to calculate the quantity of pollen load and the colourated ones (10 flowers were repeated such experiment). A blackish blue colour in the pollen load indicated the viability of pollen and no colour change or yellow-brown colour indicated the inviability of pollen. The receptivity of stigma was detected by the same MTT method (Rodriguez-Riaño and Dafni, 2000) as well.

1.3.3. Tests on Nectar Volume, Sugar Content and Amino Acid Content

20 flowers bagged before opening and 20 flowers unbagged before opening were measured via a 5μl – 10μl micropipette to measure the nectar volume every 3 hours from 7:00 to 19:00. If there was any nectar, the sugar content (%) was measured directly by a handheld refractometer; and the amino acid content was detected by the Dafni (1992) method.

1.3.4. Detection of Flower Odor

Twenty opened flowers of *P. armeniacum* and *H. beanii* were randomly selected from the population. The inflorescences were respectively put into 100 ml headspace bottles and the bottles were sealed. The odor was extracted with CAR/PDMS 75 μm (30 min), desorbed (3 min, 200 ℃), and analyzed on a Finnigan TRACE GC-MS (25 ℃, 65% humidity). The volatile components were analyzed at the Analysis Center of South China Agricultural University.

1.4. Observation on Insect Visitors and Their Visiting Behaviours

1.4.1. Observation on Insect Visitors and Their Visiting Behaviours

Ten opened flowers in population were randomly selected and marked for a 10-day observation each year from 2003 to 2005. The visiting behaviour of each insect was photographed, pictured (Canon A640, Japan) and video-recorded (Sony DCR-HCZIE, Japan) to study the pollination mechanism of *P. armeniacum*. Ten flowers in the population were randomly selected and marked each year from 2005 to 2007 for consecutive observation from flowering to being pollinated, or till finishing opening and withered. The following data were recorded: insect species for each inflorescence, their visiting frequency and the duration of each visit, and the number of flowers visited by each individual insect, in order to research the status of crossing and selfing and the mixed mating. Insects were subsequently captured for specimen evidence.

1.4.2. Test of Attraction to Pollinators by Floral Characters

In order to test the attracting effect of floral characters to pollinators, the following treatments were set: shrubberies (making up the wild habitat) were placed in a transparent glass box of 100cm × 100 cm × 80 cm into which the flowers were put, including intact flowers (for natural control), flowers with labellum removed, flowers with petals removed, flowers with sepals removed, flowers with staminode removed, flowers with stamens (anther) removed, flowers with stigma removed, and flowers with other parts removed but stigma. Five flowers were checked in each treatment, and then *Eristalis tenax, Ceratina morawitzi* and *Lasioglossum metis* (10 for each species) were put into the box. The visiting frequency and visiting behaviour of each pollinating insect to each treated flowers were observed. This experiment was repeated 15 times.

1.4.3. Test of Attraction to Pollinating Insect by Stimulating Associated Plant

The flowering status of the associated plant in population was observed, and the behaviour of visiting insects was traced, to test the mode of attracted pollinators by stimulating the associated plant. The following treatment was set: in a transparent glass box of 100 cm × 100 cm × 80 cm the following subjects were put: shrubberies, 10 flowers of *P. armeniacum*, 10 flowers of the associated plant, and pollinating insects (one species each time, 10 for each species). And then, the visiting orientation and the behaviour of pollinators were recorded, and their visiting frequency was calculated, and the result of their visiting was compared to speculate the simulating mode of *P. armeniacum*. Each experiment was repeated 10 times.

1.5. Detection of Breeding System

In order to examine the breeding system of *P. armeniacum* and the effect of self-pollination and cross-fertilization, the following treatments of artificial pollination were set up during the flowering periods from 2003 to 2006 (15 samples were treated at each year): (1) intact flowers (control) without any treatment and marked till fruited; (2) flowers removed stamens before opening and bagged; (3) flowers removed stigma before opening and bagged; (4) flowers bagged before flowering, cross-pollinated artificially after flowering, then bagged again; (5) flowers bagged before flowering, self-pollinated artificially after flowering, then bagged again; (6) flowers bagged before flowering without any further treatment till the flower withered. The fruited rate of each treatment was calculated.

1.6. Detection of Inbreeding Depression

Twenty fruits were randomly picked down from each treatment of cross-pollination and self-pollination in the detection of breeding system during the mature period of fruit in September from 2004 to 2007 (2 fruits from each population at each year). And then the quantity of seeds in each fruit was counted and all the seeds were randomly sown on artificial culture medium. The quantities of germinated seeds and un-germinated seeds were counted after germination. The degree of inbreeding depression (δ) was estimated via the differences between the average germinated rate of self-descendants (W_s) and cross-descendants (W_o), $\delta=1-W_s/W_o$. Finally, the average values of germinated rate in selfing and crossing were mated up with μ-detection to compare the difference in average germinated rate of each population.

1.7. The Failure Rate of Pollinators and the Selfing Rate

Totally there were two hundred flowers with stamens removed during budding period each year (20 flowers from each population) in the 10 populations from 2004 to 2007. They were compared with another 200 intact flowers (20 flowers from each population) for pollination observation in natural conditions. The 200 intact flowers were artificially cross-pollinated (20 for each population). Formula of calculation: the failure rate of pollinators = 1 – the natural fruited rate of intact flowers / the artificially cross-fruited rate of intact flowers; the self-pollinating rate of population = the natural fruited rate of intact flowers – the fruited rate of flowers with stamens removed; the cross-pollinating rate of population = the fruited rate of flowers with stamens removed.

2. Results and Analysis

2.1. The Morphological Structure of Flower and the Characteristics of Flowering Phenology

2.1.1. Flowering Phenology and Morphological Characteristics of Flower

The flowering period of *P. armeniacum* is from April to June and each *P. armeniacum* plant bears only 1 flower. The scape is 25 – 35 cm long; the flower is yellow-apricot; the labellum is saccate with 1 entrance and 2 exits and purple-brown spots inside; the staminode is yellow and smooth-faced and purple-brown steaks, erect above the sac opening; and the anther membrane becomes mucous in 5.5±1.5 days after flowering (n=80) (Figure 5.3. a, e). In natural condition, flowers that are not fertilized will dry out in 30.5±5.6 days after opening (n=80). After pollination the colour of flowers will become pale, the fragrance will disappear, and the perianth will completely dry out and drop in 7.6 ±1.7 days from ovary top (n=80). After 6 months of inoculation, capsules have matured and dehisced, dispersing seeds.

2.1.2. Morphological Structure of Pollen Grains and Stigma and the Relative Position of Anthers and Stigma

The stigma is oval, approximately 0.8–0.9 cm long and 0.4–0.5 cm wide with a downward receptive surface for pollen grains, and lays anterior to the basal channel of labellum. There are two stamens, with 0.3–0.4 cm filaments, respectively situated near the exit of the basal channel sides, 0.3–0.4 cm away from the stigma (Figure 5.3. a, b). Based on the SEM observation, the surface of anther membrane is sticky and the anther is full of spherical pollen grains and sticky substance (Figure 5.3. c), and the receptive surface of stigma is densely covered with papillae (Figure 5.3. d).

2.2. Flowering Phenology

2.2.1. The Quantities of Pollen Grains and Ovules and the P/O Ratio

The average quantity of pollen grains and ovules in a single flower of *P. armeniacum* were 916370 ± 19073 (n=20) and 40896 ± 7635 (n=20) respectively. The P/O ratio was 22.41 ± 2.50 (n=20).

2.2.2. The Viability of Pollen and the Receptivity of Stigma

All the samples of pollen loads and stigma appeared a coloration of blackish blue, and the colour of control groups had no change, which indicated the pollen had viability and the stigma had receptivity during flowering period and there was no anthesis of dichogamy.

2.2.3. Nectar Volume and Amino Acid Content

No nectar was detected in the flower of *P. armeniacum*, neither in bagged treatment nor in natural condition, so it was impossible to test the sugar content in nectar and amino acid content.

2.2.4. Flower Odor

Gas chromatiogram of floral odor of *P. armeniacum* and *H. beanii* refers to Figure 5.1 and Figure 5.2. The flower of *P. armeniacum* gave off a rape-like fragrance on occasion, similar to that of *H. beanii*.

2.3. Visiting Insects and Their Visiting Behaviours

2.3.1. Visiting Insects and the Mechanism of Pollination

The flower of *P. armeniacum* has no nectar and no pollens provided as rewards to the visiting insects (see 2.2.3). Its fragrance is similar to that of *H. beanii*. The purple-brown veines on staminode surface and the purple-brown spot inside the labellum (Figure 5.3. a, g) are similar to the colour of stamens of rewarding plants, which deceives pollinating insects into the saccate labellum (see 2.2.4, 2.3.2 and 2.3.3). Those insects of larger size, such as

Eristalis tenax (L.), *Ceratina morawitzi* Sickmann, *C. chinensis* (Wu), and *Heriades sauteri* Cochkerell, would land on the staminode or labellum while hovering around flowers, to forage on the veins of staminode by crawling up the staminode and then slipped into the sac.

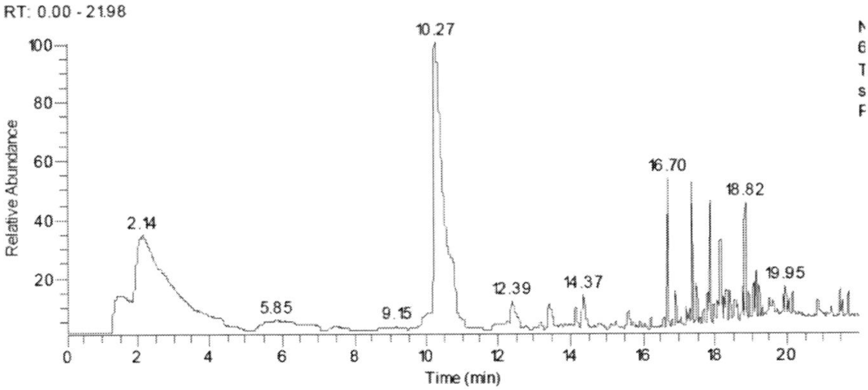

Figure 5.1. Gas chromatiogram of floral odor of *P. armeniacum*.

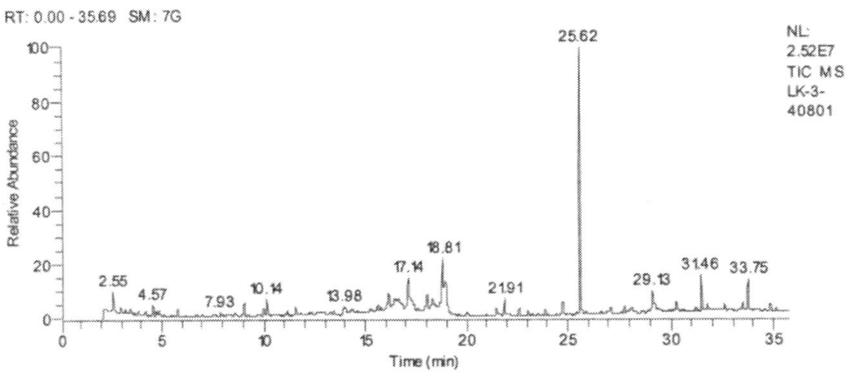

Figure 5.2. Gas chromatiogram of floral odor of *H. beanii*.

After finding no food, these insects could not retreat through the former route but only find their escape through the basal channel in the sac. When escaping they would touch the surface of stigma firstly, and sencondly the stamen, bearing pollen grains on their back. This process was repeated in another flower, namely, pollen grains from the previous flower were deposited on the stigma of latter flower, and the pollens of the latter flower were brought away, which resulted in cross-pollination (Figure 5.3. a, e, g).

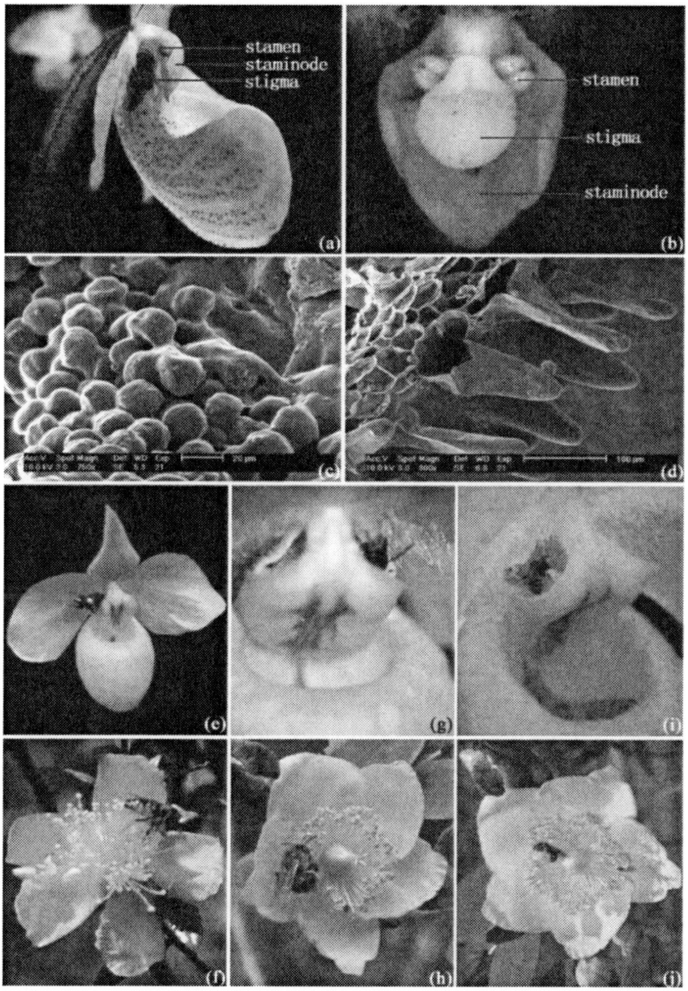

Figure 5.3. Flower morphology and pollinator behavior. (a) Longitudinal section of labellum (side view),showing the position of stigma and stamen and an *E. tenax* escaping from basal channel of labellum. (b) The stigma and stamen (back view). (c) SEM shows the surface of anther membrane and pollen grains. (d) SEM shows the papilla on the receptive surface of stigma. (e) *E. tenax* escaping from basal channel of labellum of *P. armeniacum* flower. (f) *E. tenax* licking pollen on the flower of *H. beanii.* (g) *C. morawitzi* escaping from basal channel of labellum of *P. armeniacum* flower. (h) *C. morawitzi* licking pollen on the flower of *H. beanii.* (i) *L. pronotale* crawling on the stamen of *P. armeniacum* flower. (j) *L. pronotale* licking pollen on the flower of *H. beanie.*

And insects of smaller size, such as *Lasioglossum spinodorsum*, *L. pronotale* and *L. metis*, would also land on the staminode, and then slip into the sac or on the surface of labellum to crawl into the sac, or they would fly directly into the sac while hovering around the flower. These insects would be trapped and stayed in the labellum (sometimes as long as about 30 minutes), and they moved across the stamen and the surface of stigma. When they moved on the mucous pollinium, their legs would be adhered with pollen grains, and they would deposit the pollen grains on papillae when crawling onto the stigma (because some insects have the habits of clearing up their legs before escaping), therefore self-pollination was achieved (Figure 5.3. i). It is possible that some self-pollinating insects would carry a few remained pollen grains on their legs and then visit another flower, which could cause a few ovules to be cross-fertilized; just like that cross-pollinating insects could adhere a few pollens when escaping from the basal exit and occasionally retreat back to the bottom of labellum, which could bring up a few self-fertilized ovules. The effect of such two behaviours could be counteracted and ignored.

During the 4-year observation on 40 flowers, the average visiting frequency (the summation of all visitors) was 0.11 times per 60 minutes. During the observation, there were 8 species of insects successfully visiting flowers, and 21 flowers were visited and 19 flowers were never visited. The frequency of different flowers and the same flower being visited is significantly different. As for the 21 visited flowers, there were 9 only visited by cross-pollinating insects (cross-pollination); 6 only by *Lasioglossum* spp., 2 of which were visited both at the early stage and middle stage of flowering period (self-pollination); 2 were visited at late stage of flowering period (delayed selfing); 6 were visited by two species (competing within-flower selfing), 4 of which were visited firstly by cross-pollinating insects and then soon by self-pollinating insects before the perianth fell down, and the other 2 were visited by self-pollinating insects firstly and then by cross-pollinating insects before the perianth fell down (prior selfing).

The colour of flowers that were successfully fertilized gradually became pale, the veins of staminode and fragrance disappeared, and the perianth would dry out and fall down 7 – 8 days later. No insects were found foraging for pollen grains on the stamens, nor did they oviposite or copulate on the flowers.

2.3.2. The Attraction of Floral Organs to Pollinating Insects

Eristalis tenax, C. morawitzi and *L. metis* were put into the glass box to test the visiting frequency of different treatments. It was shown that the results of treatments with flowers removed labellum, stigma, petals and stamens all had no difference with control, which indicated that the attraction to *E. tenax, C. morawitzi* or *L. metis* by removing labellum, stamens, petals and stigma did not obviously decrease, which showed that pollinators could not differentiate flowers without stamens. And the differences of treatments with flowers removed staminode and those removed other floral parts but only remained stigma were significant from that of control, indicating that stigma had no attractive effect on pollinating insects and the existence of staminode had extremely obvious attraction to pollinating insects.

2.3.3. Simulation of Associated Plant

Hypericum beanii is the only associated plant that has the same flowering period as *P. armeniacum*. *Hypericum beanii* has yellow-apricot petals and light purple-brown stamens and gives off a slight rape-like fragrance. Pollinators of *P. armeniacum* all foraged on the pollen grains of *H. beanii* for food (rewarding plant for pollinating insects) (Figure 5.3. f, h, j). The result of tests on visiting frequency of pollinating insects (*E. tenax*, *C. morawitzi*, *L. metis*) when placed with *H. beanii* and *P. armeniacum* together is shown as Figure 5.4: there is no difference in the visiting frequency of *P. armeniacum* and *H. beanii* by *E. tenax*, *C. morawitzi, L. metis* (t=0.1035, $d.f.$=18, P>0.05; t=0.39417, $d.f.$=18, P>0.05; t=0.32769, $d.f.$=18, P>0.05).

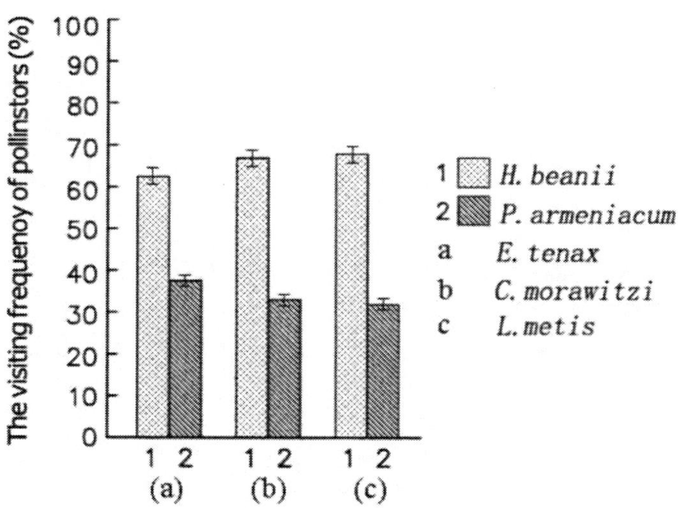

Figure 5.4. The visiting frequency of pollinator in *P. armeniacum* in pair with co-flower (mean±s.d., n=20).

Pollinating insects have an inclination of visiting the rewarding plants more often. *Paphiopedilum armeniacum* provides no rewards for pollinating visitors, but its fragrance, coloration and floral structure are similar to that of *H. beanii*, which confuses insects that are about to forage for the pollen grains of *H. beanii*. *Paphiopedilum armeniacum* simulates the flower of a rewarding plant and designs a floral structure for cross-pollination and self-pollination, which deceives pollinators into a trap to achieve pollination.

2.4. Detection on Breeding System

The result of detection on breeding system demonstrated that there is no significant difference between the fruited rates of artificial self-pollination and artificial cross-pollination (t=0, *d.f.* =118, P>0.05), indicating *P. armeniacum* is compatible in both self-pollination and cross-pollination. The fruited rates of natural compared and treatment of bagged before opening (33.16 ± 10.87 % vs 0 %, respectively) are obviously different (t=7.211, *d.f.* =118, P<0.01), indicating that *P. armeniacum* cannot achieve sexual production via autonomous self-pollination in natural conditions. The flowers with stamens and pistils removed set no fruit, indicating *P. armeniacum* cannot produce asexual seeds. There are significant differences in control and artificial self-pollination and cross-pollination (t=9.090, *d.f.* =118, P<0.01 and t=9.090, *d.f.* =118, P<0.01), indicating that there are insufficient pollinators to pollinate all flowers in natural conditions (see 2.6).

2.5. Detection of Inbreeding Depression

After calculation, the quantity of seeds and the germinated rate of each fruit are shown in Table 5.1.

The results show that there is no significant difference in quantity of seeds between self-pollination and cross-pollination (u=1.0105, n=80, P>0.05), but there is obvious difference in germinated rates of self-pollination and cross-pollination (u=166.9840, n=80, P<0.01), which indicates that self-pollination can increase the quantity of fruits of population and there is no seed discounting but the obviously different germinated rate shows there is serious inbreeding depression: $\delta=1-W_s/W_o=1-3.55/29.33=0.88$.

Table 5.1. The seed quantity and the germinated rate (GR) of self- and cross-pollination

Year	Cross-pollination		Self-pollination		ID (δ)
	Seed quantity	GR (W_o)	Seed quantity	GR (W_s)	
2004	38953±2136	30.09±2.30	38821±1920	3.53±0.33	0.88
2005	38475±612	29.05±0.72	38336±611	3.48±0.30	0.88
2006	39224±938	29.22±0.64	39076±964	3.54±0.30	0.88
2007	39448±1026	28.99±0.77	39473±943	3.67±0.24	0.87
$\bar{x}$	39025±1328	29.33±1.35	38926±1251	3.55±0.29	0.88

Notes: GR for germinated rate; ID for inbreeding depression.

2.6. The Failure Rate of Pollinator and the Selfing Rate

The failure rate of pollinator and the selfing rate of each population are shown as Table 5.2.

Based on the results in Table 5.1, cross-pollinating rate of fruit set (average) = FR of removed stamen/FR of natural=20.75/37.63=55.14%; Self-pollinating rate of fruit set (average)= 1– Cross-pollinating rate of fruit set (average)= 1–55.14%=44.86%.

As there is competing selfing in *P. armeniacum*, the quantity of seeds in each capsule and the germinated rate of offsprings from selfing and crossing of plants in each population were compared to estimate the degree of inbreeding depression.

Table 5.2. The lacking rate (LR) of pollinator and the self-rate (SR) in population

Year	Natural			Removed Stamen			Artificial Cross			LR	SR
	Quantity	FQ	FR(%)	Quantity	FQ	FR(%)	Quantity	FQ	FR(%)		
2004	200	49	24.50	200	11	5.50	200	198	99.00	75.50	19.00
2005	200	96	48.00	200	62	31.00	200	197	98.50	51.27	17.00
2006	200	67	33.50	200	40	20.00	200	196	98.00	65.82	13.50
2007	200	89	44.50	200	53	26.50	200	200	100	55.05	18.00
$\bar{x}$	200	75.25	37.63	200	41.50	20.75	200	197.75	98.88	61.91	16.88
SD	0.00	18.55	9.28	0.00	19.27	9.63	0.00	1.48	0.74	9.49	2.07

In the 40 compared groups over four years, there is no statistical difference in the quantity of seeds between groups of self-pollination and cross-pollination (u=1.0105, n=80, P>0.05). As for the germinated rate of seeds in all 10 populations, there is an obvious difference between the groups of self-pollination and cross-pollination (u=166.9840, n=80, P<0.01), which shows a high level of inbreeding depression (Table 5.1), with value more than 0.5 of competing selfing. Though the occurrence of selfing in *P. armeniacum* is slightly delayed and there is delayed selfing, the costs of pollen discounting and seed discounting are rather high in environment of sufficient cross-pollinators.

The variation of pollinating environment (visiting insects) had been quantified in the first 3 years of this experiment. The successful fruited rates of flowers in natural control, removed stamens, and of the intact cross-pollinated flowers were compared (Figure 5.5).

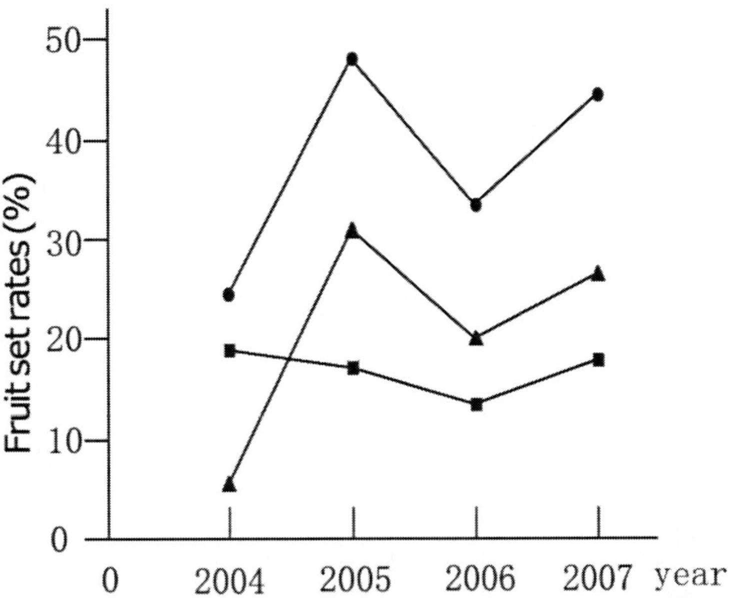

Figure 5.5. Fruit set rates (%) of nature, cross- and self-pollination. Fruit set rates in nature (—●—) is the result of crossing (—▲—) and selfing (—■—) of pollinators. Its saturation shows lacking pollinators leaves many flowers unfertilized so that selfing has reproductive assurance; the difference between selfing rate and crossing rate exists, which changes with years and results in mixed mating and its change.

The results show the difference of visiting behaviour by cross-pollinators and self-pollinators during flowering period and among flowering periods is significant, and the low visiting rate of cross-pollinator has brought about an obvious drop on the fruited rate of flowers with stamens removed. More importantly, the fruited rate of flowers with stamens removed and that of intact ones were compared, and it shows that self-pollination via pollinating agents can increase the fruited rate in condition of no cross-pollinators or insufficient cross-pollinators. The phenomenon of mass flowers un-pollinated during flowering period indicates that there is a shortage of both cross-pollinators and self-pollinators. Therefore, the ecological environment of reproductive assurance has existed in communities of *P. armeniacum* all the time.

3. Discussions

Selfing and mixed mating are common in plants (Barrett, 2002; Lloyd and Schoen, 1992). The evolution of selfing is restricted by the cost of fitness of selfing descendant when compared to that of crossing descendant (Lloyd and Schoen, 1992; Jarne and Charlesworth, 1993; Holsinger, 1991). When crossing is impossible to carry out, self-pollination has an adaptive advantage in aspect of providing reproductive assurance (Darwin, 1876; Stebbins, 1974; Lloyd, 1992). The changeful pollinating environment is favorable for the occurrence of selfing and the formation of mixed mating (Schoen and Brown, 1991; Kalisz et al., 2004). Though the reproductive assurance of selfing and the evolutionary hypotheses of mixed mating are accepted widely and they have been testified by some experiments of autonomous selfing (within flower) without pollinating agents, they have not been testified by experiments of selfing (within flower) completed by pollinating insects (Kalisz et al., 2004). The quantity of various pollinators and the time needed for pollen grains need to reach the pistils in natural condition were compared in order to quantify prior selfing, competing selfing, and delayed selfing (Lloyd and Schoen, 1992; Lloyd, 1992) of self-pollination via pollinating agents. It has not yet been forecasted that when self-pollination of a flower happens will provide reproductive assurance. Theoretically, self-pollination that happens in early stage of flowering period (prior selfing) is unfavorable, but if there is no opportunity of crossing, then this self-pollination is beneficial as well as that happens in late stage of flowering period (delayed selfing). It is because that flowers have not been visited by cross-pollinators have no discounting cost no matter when the self-pollination happens. The competing selfing of a flower results in two possible situations: if self-pollination occurs before cross-pollination so that self-pollination is of disadvantage; and if self-pollination occurs after cross-pollination so that self-pollination will only be beneficial if external pollen is insufficient, owning to the low P/O value and the undefined quantity of pollen carried by pollinators. The relation between the selfing rate of *P. armeniacum* and the crossing rate is not like that between autonomous selfing and crossing which one falls as the other rises (Kalisz et al., 2004). This reflects the randomicity of the visiting of pollinators, that is, though the visiting frequency of cross-pollinators is low, and the rate of self-pollination does not always increase, the percentage of effect of reproductive assurance in populations will be increased. If both the self-pollinators and cross-pollinators visit the same flower synchronously (during the receptive time after stigma receives pollen grains), the production of seeds of both this flower and population is the outcome of the co-effect of crossing and selfing by pollinators. If cross-pollinators are abundant, all of the above-listed self-pollination should pay the costs of seed discounting and pollen discounting no matter how many the self-pollinators are and such kind of selfing is unfavourable. In reverse, when cross-pollinators are deficient but self-pollinators are sufficient, the annual production of seeds will be the mix of selfing and crossing despite the high degree of inbreeding depression.

This study has detected the relation between the visiting rate and the selfing rate brought by self-pollinators and cross-pollinators from 2004 to 2007, based on the estimate of 10 populations of inbreeding depression. And the fruited rate of each population is between 20% and 50% (Liu et al., 2006), which reflects the fluctuating attribute of visiting behaviour of pollinators. In order to quantify the visiting rate of pollinators, the non-fruited proportion of flowers that were removed stamens in natural control and the non-fruited proportion of artificially pollinated flowers were compared. During these 4 years, only a portion of flowers in population had been visited and fruited no matter how the self-pollinators or cross-pollinators changed, and the selfing rate significantly changed along with the change of population and year (Figure 5.5). Our data shows that even though the annual quantity of flowers visited by cross-pollinators and self-pollinators is different, more than 50% of flowers in the population still remain un-pollinated and the crossing rate is always low, leaving large room for the rate of pollination (including the selfing rate). Therefore there are ecological factors suited for the co-existence of self-pollination and cross-pollination, even though there is potential of competing selfing. Our data indicates that flowers which seem to have a strict structure of crossing also have a mechanism of selfing, and mechanically connect the reproductive assurance and the selfing of pollinating agents in the changeable condition that the seed production of mixed mating system could be increased or balanced.

Our data provides a full-scale demonstration for reproductive assurance of selfing via pollinating agents in the circumstance of unsaturated pollinators and provides an experimental demonstration for the selfing rate of population in the circumstance of unstable pollinators. It not only reveals a reproductive countermeasure of selfing during the evolution of mating system in *P. armeniacum*, but also extends the meaning of our study into the aspect of evolution of mixed mating system. In field populations, the instability of pollinators is normal (Kalisz et al., 2004; Liu et al., 2006; Burd, 1994), which stimulates flowers of cross-pollination to develop mechanisms so that it can attract self-pollinators when there is a shortage of cross-pollinators and maintain a regular rate of selfing with no need of evolving autonomous self-pollination. On this account, it acts as a response to the environment of undefined pollinators and consequently protects the advantage of cross-pollination and the reproductive assurance of self-pollination (Liu et al., 2006), to support the stability of a mixed mating system with a fitness of transitive selfing rate less than 0.25 (Johnston, 1998; Barrett et al., 1996). The advantages of selfing and crossing in this system will fluctuate along with the changes of variety, quantity, and behaviour of pollinators; therefore, plants may evolve from ecological selfing to genetic selfing.

REFERENCES

Arroyo, M. T., Armesto, K. J., Villagran, C. (1981). Plant phonological patterns in the high Andean cordillera of central Chile. *Journal of Ecology*, *69*, 205–223.

Barrett, S. C. H. (2002). The evolution of plant sexual diversity. *Nature Reviews Genetics*, *3*, 274–284.

Barrett, S. C. H. (2003). Mating strategies in flowering plants: the outcrossing-selfing paradigm and beyond. *Philosophical Transactions of the Royal Society of London Series B-Biological Sciences*, *358*, 991–1004.

Barrett, S. C. H., Harder, L. D., Worley, A. C. (1996). The comparative biology of pollination and mating in flowering plants. *Philosophical Transactions of the Royal Society of London Series B-Biological Sciences, 351*, 1271–1280.

Burd, M. (1994). Bateman's principle and plant reproduction: the role of pollen limitation in fruit and seed set. *The Botanical Review*, *60*, 83–139.

Darwin, C. (1876). *The Effects of Cross and Self Fertilization in the Vegetable Kingdom*. London: John Murray.

Kalisz, S., Vogler, D., Fails, B., et al. (1999). The mechanism of delayed selfing in *Collinsia verna* (Scrophulariaceae). *American Journal of Botany*, *86*, 1239–1247.

Fausto, J. A. J., Eckhart, E. V., Geber, M. A. (2001). Reproductive assurance and the evolutionary ecology of self-pollination in *Clarkia xantiana* (Onagraceae). *American Journal of Botany*, *88*, 1794–1800.

Goodwillie, C., Kalisz, S., Eckert, C. G. (2005). The evolutionary enigma of mixed mating systems in plants: Occurrence, theoretical explanations, and empirical evidence. *Annual Review of Ecology Evolution and Systematics, 36*, 47–79.

Herlihy, C. R. and Eckert, C. G. (2002). Genetic cost of reproductive assurance in a self-fertilizing plant. *Nature*, *416*, 320–323.

Holsinger, K. E. (1991). Mass-action models of plant mating systems: the evolutionary stability of mixed mating systems. *American Naturalist*, *138*, 606–622.

Holsinger, K. E. (1996). Pollination biology and the evolution of mating systems in flowering plants. *Evolutionary Biology*, *29*,107–149.

Jain, S. K. (1976). The evolution of inbreeding in plants. *Annual Review of Ecology and Systematics*, *7*, 469–495.

Jarne, P. and Charlesworth, D. (1993). The evolution of the selfing rate in functionally hermaphroditic plants and animals. *Annual Review of Ecology and Systematics*, *24*, 441–466.

Johnston, M. O. (1998). Evolution of intermediate selfing rates in plants: pollination ecology versus deleterious mutations. *Genetica*, *102*/*103*, 267–278.

Kalisz, S., Vogler, D. W., Hanley, K. M. (2004). Context-dependent autonomous self-fertilization yields reproductive assurance and mixed mating. *Nature*, *430*, 884–887.

Liu, K. W., Liu, Z. J., Huang, L. Q., et al. (2006). Self-fertilization strategy in an orchid. *Nature*, *441*, 945–946.

Liu, Z .J., Chen, L. J., Lei, S. P., et al. (2007). The reproduction strategy of *Trias verrucosa* from China. *Acta Ecologica Sinica*, *27* (11), 4460–4468.

Lloyd, D. G. and Schoen, D. J. (1992). Self- and cross-fertilization in plants. I. Functional dimensions. *International Journal of Plant Sciences*, *153*, 358–369.

Lloyd, D. G. (1980). Benefits and handicaps of sexual reproduction. *Evolution, 13*, 69–111.

Lloyd, D. G. (1992). Self- and cross-fertilization in plants. II. The selection of self-fertilization. *International Journal of Plant Science*, *153*, 370–380.

Rodriguez-Riaño, T. and Dafni, A. (2000). A new procedure to assess pollen viability. *Sex Plant Reprod*, *12*, 241–244.

Schoen D. J., Morgan M. T., Bataillon T. (1996). How does self-pollination evolve? Inferences from floral ecology and molecular genetic variation. *Philosophical Transactions of the Royal Society of London Series B-Biological Sciences, 351*, 1281–1290.

Schoen, D. J. and Brown, A. H. D. (1991). Whole- and within-flower self-pollination in *Glycine argyrea* and *G. clandestine* and the evolution of autogamy. *Evolution*, *45*, 1651–1665.

Stebbins, G. L. (1974). *Flowering Plants: Evolution Above the Species Level 52.* Cambridge: Belknap.

Totland, O. and Sottocornola, M. (2001). Pollen limitation of reproductive success in two sympatric alpine willows (Salicaceae) with contrasting pollination strategies. *American Journal of Botany*, *88*, 1011–1015.

Vasek, F. C. and Weng, V. (1988). Breeding systems of *Clarkia* sect. *Phaeostoma* (Onagraceae): I. Pollen-ovule ratios. *Systematic Botany*, *13*, 336–350.

Chapter 6

CONCLUSION AND CONSERVATION

ABSTRACT

This chapter summarizes the studies of the above five chapters. It is believed that *Paphiopedilum armeniacum* makes the best of environment to serve propagation so that it gains relative competitive advantage, which represents that the change of adaptability related to the life history evolution may be the essential condition of its being an invasive species. It indicates that plant not only has the ability to develop crossing but also has the ability to promote selfing and employ selfing, and the sexual change may be the product from the tradeoff in life history of new habitat invasion and habitat fluctuation. The revelation gained from the reproductive strategy of *P. armeniacum* is that, there is no plant can use any environment, and few environments are not used by plant, and then gives insight into the protection of this species.

The habitat of *P. armeniacum* is unique. Based on the selective result of habitat of its relative species, *P. armeniacum* has its distribution in West Nushan Mountain and the adjoining region in Yunnan, which shows that the main motive of differentiation in angiosperm is probably many ecological opportunities that can be exploited. The change of habitat may impel the change of breeding system and this change is always accompanied by the change of structure of conformation. The result of observation on the breeding method of *P. armeniacum* shows, the orchid has both asexual reproduction and sexual reproduction, and the sexual reproduction builds and maintains the mixed mating system making use of pollinating insects, which is obviously different from the special mechanism of pollination in other orchids.

Paphiopedilum armeniacum makes good use of the environmental resources to serve its breeding in order to attain the relative dominance in competition. Pollinators can accelerate the separation of breeding among populations by its structure or the mechanism of behaviour, resulting in the high diversification of angiosperm. The existence of *P. armeniacum* in this habitat and its relation with the pollinating insects facilitate the differentiation between *P. armeniacum* and its relative species, which confirms that the frequent change of environment and bad conditions are propitious for the appearance of new species. *Paphiopedilum armeniacum* has achieved the separation of breeding via the adaptation to various kinds of pollination and the morphological differentiation of diffusing agents of seeds, to create new species. During this process, it has carried out different patterns of growth and life history and shaped its new breeding strategy.

Paphiopedilum armeniacum is often distributed in a group, which is primarily the result of its own seed germination and the asexual reproduction. During investigation in the field, many seedlings were discovered nearby mature plants of *P. armeniacum*. These seedlings could become mature after growing 3 years, and individuals were relatives. As this species carries out a strategy of mixed mating to maintain its adaptability at the time it exploites new habitat, and its growth, development and breeding all become mature rapidly, and the seeds are often distributed around their mother genets as well. It will not cause competition among sibling ramets; on the contrary, it will increase the opportunity for pollination among relatives. The result of the experiment indicates that individuals of serried distribution can raise the selfing rate or the rate of inbreeding, which can maintain its occupation in the new habitat. It will have adjustment in adaptability of the resource allocation among growing, survival and breeding (Willis et al., 2000; Thebaud and Simberloff, 2001). The adjustment of adaptability of *P. armeniacum* in its life history further strengthens its ability for invasion in the environment. The results of observation manifest that the change of adaptability relative to its life history in the evolutionary theory of life history may be the requirement of its ability of invasion.

In condition of insufficient pollen grains owing to very low density of population or lack of pollinating agents, selfing is also selected to ensure the fertilization of the ovule, so as to bring out the effect of breeding assurance, even though inbreeding depression is very strong. The rates of selfing or crossing or mixed mating in *P. armeniacum* is never saturated and the reproductive assurance of this species exists all the time. Thus, it is just for its reproductive assurance that it can survive and multiply via the evolution of adaptability of selfing in the condition of bad habitat, and it is also a reason why *P. armeniacum* carries out the current ecological strategy. The plant of *P. armeniacum* only opens one flower and the flower has strict structure of fertilization; however, it possesses strategy of selfing, which indicates that this plant is not only provided with the ability of developing crossing but also with the ability of promoting and demanding selfing (Liu et al., 2008), moreover it gives an explanation to the breeding strategy of selfing or mixed mating that is used by most plants in nature.

Sexual reproduction is possessed by nearly all eukaryotes and it is a rather primitive character, which is possibly helpful for the reproduction and long-term survival of living creatures. Variability of heredity has provided potential power of evolutionary for the offsprings of sexual reproduction: when there are changes in the environment, sexual offspring can adapt to the new environment, whereas offspring of asexual reproduction can hardly make rapid adaptive response to a new environment. However, asexual offspring do not need to expend about half of its energy in producing male offspring (or pollen grains) and it can pass all genes down to the next generation (sexual offspring can only pass half of the genes), so the asexual offspring have more obvious advantages than sexual offspring in the competition of survival regarding population statistics (Zhang, 2004). Therefore, gender is a kind of paradox during the process of evolution: the disadvantages of sexual reproduction in evolution are much more obvious than its advantages, but actually, it exists universally in the natural world. It is just like the fact that selfing can bring inbreeding depression but it is still used by most plants. Then, what makes sexual reproduction triumph in the evolutionary competition? Actually, from aspects of the geographic distribution and reproductive patterns of *P. armeniacum*, there is another explanation for the evolution of gender, that is, it's the product of tradeoff between the invasion of new habitat and the life history of fluctuating habitat. Selfing is an intermediate stage between out-crossing and asexual reproduction that possesses the merits of both reproduction.

Avoidance of selfing was used to explain nearly all the reproductive characters of plants in the past. An important topic of the discussion among biologists of evolution is to clarify the mating strategy of flowering plants, the origin of diversity of gender and the selective force of its maintenance. They not only need to understand how they (plants) try to avoid selfing and the disturbance from female and male functions, but also need to find out how they promote selfing and harmonize the expression of female and male functions (Zhang, 2004). In fact, it is possibly underestimated the significance of adaptability of how plants promote selfing and the diversity of selfing, as well as how selfing affects the new habitat (Darwin, 1876). It's true that selfing or inbreeding can reduce the diversity of heredity, but this reduction will not necessarily result in the bankruptcy of adaptive potential. Generally, inbreeding depression is not of an unvarying value, and it will change according to the degree of asperity in the environment. And it will show a neutral attribute or a slight disadvantageous character in the environment of preponderance, and it will become harmful in an execrably threatening environment (Dudash, 1990; Wolfe, 1993; Hauser and Loeschcke, 1996; Cheptou et al., 2000a; Cheptou et al., 2000b). *Paphiopedilum armeniacum* shows a severe inbreeding depression from the aspect of seed breeding, but it behaves badly in a superior habitat yet grows very well in a bad habitat. It's probably because the breeding system of *P. armeniacum* has lost the potential of adaptability in a new superior environment after adapting to the extreme ecological environment; however, *P. armeniacum* shows a very good ability of invading a new bad habitat. So, it needs to re-survey and judge the standard of goodness and badness of the environment in every point of view from the adaptability aspect. As for *P. armeniacum*, to a certain extent, a 'bad environment' is a kind of superior environment. And clonal reproduction and selfing both have dominance in such kind of habitat, for survival is much more important than any other kind of dominance. The revelation of inbreeding strategy of *P. armeniacum* is that the understanding of its function should not be separated from its background, at least from the aspect of sexual reproduction of plants (Herlihy and Eckert, 2002; Qu et al., 2007). In the world, there is no plant that can make use of any environment and there are few environments that are not used by plants.

Paphiopedilum armeniacum, since its first discovery, has spread to many places in the world by different means. There are well-cultivated plants of *P. armeniacum* appearing in international orchid exhibitions (Chen and Luo, 2003), which partly shows that *P. armeniacum* can be artificially cultivated in *ex situ*. Based on the test result of *ex situ* conservation, it is known that this species absolutely can be *ex situ* conserved, and from the test of reintroduction, those plants of asexual reproduction are well seasoned with the original habitat. Therefore, it can be concluded that there will be no technical obstacle for *ex situ* conservation and reintroduction of *P. armeniacum*, and the methods are simple and easy to handle, and less expensive as well. So it can rejuvenate the wild communities of *P. armeniacum* via large scale artificial multiplication and reintroduction of progenitive plants in the original habitat. Certainly the transmissibility of a community of *P. armeniacum* should be taken into consideration in order to preserve the diversity of resources of this species, so that *P. armeniacum* will not become extinct because of its own biological defects. At present, the threats that *P. armeniacum* has encountered, besides the excessive gathering, are mainly from over-sabotage of its habitat from human-made destruction, which entirely deprives it of space for growth or results in irreversible changes of its habitat. This is the main reason that leads to the extinction of *P. armeniacum*.

Paphiopedilum armeniacum is a species with a narrow distribution in China and its "hot spot" in Nushan Mountain. Since it has the outward diffusibility and the invading ability, *P. armeniacum* could be kept in a certain amount in brushes of *Q. gilliana* and likewise in a limestone area, while it would soon invade the nearby habitat and massively multiply after a resuming of small habitats such as plantations returning to forests and earth no longer dug or washed out by rain. Therefore, conservation of *P. armeniacum* should start with the protection of its own habitat: to set up the conservation sites, to establish mother communities, to implement close protection, to eliminate the threats to *P. armeniacum* posed by human activities of economic trade to maintain habitat stability, to meet the growing condition of rocks in the limestone area, to protect environment completely for pollinating insects, etc. At present, it is necessary to carry out the *ex situ* conservation and the artificial reproduction of *P. armeniacum* as soon as possible, to store up mass seedlings of artificial reproduction prepared for wild re-introduction, to renew this endangered species and to make it thriving and prosperous.

REFERENCES

Chen, S. C. and Luo, Y. B. (2003). Advances in some plant groups in China I. a retrospect and prospect of Orchidology in China. *Acta Botanica Sinica*, *45*, 2–20.

Cheptou, P. O., Berger, A., Blanchard, A., et al. (2000a). The effect of drought stress on inbreeding depress in four populations of the Mediterranean outcrossing plant *Crepis sancta* (Asteraceae). *Heredity*, *85*, 294–302.

Cheptou, P. O., Imbert, E., Lepart, J., et al. (2000b). Effects of competition on lifetime estimates of inbreeding depression in the outcrossing plant *Crepis sancta* (Asteraceae). *Journal of Evolutionary Biology*, *13*, 522–531.

Darwin, C. (1876). *The Effects of Cross and Self Fertilization in the Vegetable Kingdom*. London: John Murray.

Dudash, M. R. (1990). Relative fitness of selfed and outcrossed progeny in a self-incompatible, protandrous species, *Sabatia angularis* L. (Gentianaceae): a comparison in three environments. *Evolution*, *44*, 1129–1139.

Hauser, T. P. and Loeschcke, V. (1996). Drought stress and inbreeding depression in *Lychnis flos-cuculi* (Caryophyllaceae). *Evolution*, *50*, 1119–1126.

Herlihy, C. R. and Eckert, C. G. (2002). Genetic cost of reproductive assurance in a self-fertilizing plant. *Nature*, *416*, 320–323.

Liu, Z. J., Chen, L. J., Liu, K. W., et al. (2008). *Chenorchis*: a new orchid genus and its eco-strategy of ant pollination. *Acta Ecologica Sinica*, *26* (6), 2433–2444.

Qu, R., Li, X., Luo, Y. et al. (2007). Wind-dragged corolla enhances self-pollination: A new mechanism of delayed self-pollination. *Annals of Botany*, *100* (6), 1155–1164.

Thebaud C. and Simberloff D. (2001). Are plants really larger in their introduced ranges? *American Naturalist*, *157*, 231–236.

Willis, A. J., Memmott, J., Forrester, R. I. (2000). Is there evidence for the post-invasion evolution of increased size among invasive plant species? *Ecology Letters*, *3*, 275–283.

Wolfe, L. M. (1993). Inbreeding depression in *Hydrophyllum appendiculatum*: role of maternal effects. Crowding, and parental mating history. *Evolution*, *47*, 374–386.

Zhang, D. Y., ed. (2004). *Ecology of Evolution and Reproduction on Plant Life History*. Beijing: Science Press, 1–421.

ACKNOWLEDGMENTS

We greatly appreciate the financial support from the State Forestry Administration, P. R. China (No.2005–122) and the Science and Technology Project of Shenzhen, China (No.2006–464); and we are grateful to Professor Lai Qiang HUANG from Graduate School at Shenzhen, Tsinghua University for his experimental guidance. We give our deep gratitude to Xiao Chun SHI, Ke Wei LIU, Li Qiang LI, Si Peng LEI, Yu Ting ZHANG, Xu Hui CHEN and Yu Yun ZHENG for their excellent experimental work.

INDEX

A

B

C

D

E

F

G

H

I

L

M

N

O

P

Q

R

S

T

V

W